FLORA ZAMBESIACA

Flora terrarum Zambesii aquis conjunctarum

VOLUME TWELVE: PART FOUR

FLORA ZAMBESIACA

MOZAMBIQUE

MALAWI, ZAMBIA, ZIMBABWE

BOTSWANA

VOLUME TWELVE: PART FOUR

Edited by
G. V. POPE

on behalf of the Editorial Board:

G. Ll. LUCAS
Royal Botanic Gardens, Kew

I. MOREIRA
*Centro de Botânica, Instituto de Investigação
Científica Tropical, Lisboa*

G. V. POPE
Royal Botanic Gardens, Kew

Published by the Managing Committee on behalf of
the contributors to Flora Zambesiaca
1993

Typeset at the Royal Botanic Gardens, Kew, by
Pam Arnold, Christine Beard, Dominica Costello,
Margaret Newman, Pam Rosen and Helen Ward

Printed in Great Britain by
Whitstable Litho Printers Ltd., Whitstable, Kent.

ISBN   0 947643 57 5

CONTENTS

FAMILY INCLUDED IN VOLUME XII, PART 4

191. IRIDACEAE

LIST OF NEW NAMES PUBLISHED IN THIS WORK

Acknowledgement.

Grants from the United States National Science Foundation (most recently BSR 89-06300), and the National Geographic Society (3748-88), which made possible the completion of this treatment are gratefully acknowledged by the author.

191. IRIDACEAE

By P. Goldblatt

Perennial herbs, evergreen or dying back each year to a persistent stock, with rhizomes bulbs or corms, rarely annual herbs, or shrubs with a woody caudex. Leaves basal and cauline; sometimes the lower 2–3 membranous below, entirely sheathing, not reaching much above the ground (thus cataphylls); foliage leaves with open or closed sheaths, the blades mostly distichous, usually equitant and ensiform, parallel-veined, plane, plicate or rarely terete; or leaves bifacial and channelled to flat in a few genera. Flowering stems aerial or subterranean, simple or branched, terete, angled or winged. Inflorescence either composed of umbellate clusters (rhipidia) enclosed in opposed leafy to dry bracts (spathes) with flowers usually pedicellate (to more or less sessile) and each subtended by one bract; or inflorescence spicate with flowers sessile, each subtended by two opposed bracts; or occasionally flowers solitary. Flowers hermaphrodite, with a petaloid perianth of two equal or unequal whorls of tepals (rarely one whorl absent), actinomorphic or zygomorphic. Tepals usually large and showy, free virtually to the base, or united in a perianth tube. Stamens 3, inserted at the base of the outer tepals, or in the perianth tube, symmetrically disposed or unilateral and arcuate (or declinate); filaments filiform, free or partly to completely united; anthers 2-thecous, extrorse, usually dehiscing longitudinally. Ovary inferior (but superior in the Tasmanian *Isophysis*), trilocular with axile placentation (rarely unilocular with parietal placentation), ovules usually anatropous, many to few; style filiform, usually 3-branched, sometimes simple or 3-lobed, style branches either filiform or distally expanded, sometimes each divided in the upper half, stigmatic towards the apices, or the branches thickened or flattened and petaloid, the stigmas then abaxial below the apices. Fruit a loculicidal capsule, rarely indehiscent; seeds globose to angular or discoid, sometimes broadly winged, usually dry (rarely fleshy), endosperm hard, with reserves of hemicellulose, oil and protein, embryo small.

A family of c. 80 genera and 1700 species, more or less worldwide in distribution, but rare in tropical lowlands. It is best represented in southern Africa. The family is currently divided into four subfamilies [Goldblatt in Ann. Missouri Bot. Gard. 77: 607, (1990)]. Subfamily *Isophysidoideae* Takhtajan is monotypic comprising the Tasmanian *Isophysis* T. Moore with a superior ovary. The subfamilies *Nivenioideae* Schulze ex Goldblatt, *Iridoideae* and *Ixioideae* Klatt are represented in the Flora Zambesiaca area.

The family is of considerable economic importance in horticulture and the cut flower industry, especially the genera *Iris*, *Gladiolus* and *Freesia*. Several other genera (*Dietes*, *Crocus*, *Watsonia*) are cultivated in gardens in both tropical and temperate areas. *Moraea* and *Homeria* are poisonous, and are significant economically in cattle and sheep raising areas, notably in southern Africa. Some genera are of importance in traditional medicine, especially *Gladiolus*. Corms of several species of *Lapeirousia* and *Gladiolus* are eaten locally. The corms were an important source of food for humans in prehistoric time.

Key to the genera

1. Stamens opposite and appressed to the style branches; flowers actinomorphic, sometimes *Iris*-like with petaloid style branches with paired terminal crests; leaves sometimes bifacial and channelled or unifacial and equitant - - - - - - - - - - 2
   – Stamens alternating with or opposite the style branches, but never appressed to them; flowers zygomorphic, or actinomorphic but then not *Iris*-like, the style filiform and simple or terminating in short lobes or filiform branches; leaves unifacial and equitant, sometimes terete or almost so - - - - - - - - - - - - - - 6
2. Flowers not *Iris*-like but more or less stellate with tepals spreading or cupped below; style branches not petaloid, style branch crests if present inconspicuous or plumose-frimbriate; both inner and outer tepals with fairly inconspicuous nectar guides - - - - - 3
   – Flowers *Iris*-like; style branches petaloid, terminating in paired crests with entire margins; only outer tepals with a nectar guide at the base of the limb - - - - - - 4

3. Foliage leaf usually solitary, bifacial and channelled; style branches entire or divided apically into 2 short lobes   -   -   -   -   -   -   -   -   -   -   -   -   -   **6. Homeria**
  – Foliage leaves usually 3 or more in a distichous fan (or leaf blades not present at anthesis), unifacial and equitant; style branches terminating in plumose-fimbriate appendages
  3. Ferraria
4. Plants evergreen with creeping rhizomes; aerial parts persisting for more than one year; leaves unifacial, equitant   -   -   -   -   -   -   -   -   -   -   -   -   -   **2. Dietes**
  – Plants dying back each year to persistent corms; leaves bifacial, channelled, rarely terete   5
5. Ovary globose to oblong, long-pedicellate and exserted or included near the top of the spathes; overy fertile throughout, thus not sterile and tubular above   -   -   -   **4. Moraea**
  – Ovary elongate-cylindric, more or less sessile and included within the spathes; ovary fertile in the lower third, tapering above into a sterile tube-like beak   -   -   **5. Gynandriris**
6. Tepals free virtually to the base; flowers arranged in clusters of 2 or more in biseriate rhipidia, these stalked or sessile; perianth blue, fugaceous, open in the morning, collapsing spirally and deliquescing in the early afternoon; rootstock a rhizome   -   -   -   -   **1. Aristea**
  – Tepals clearly united in a perianth tube; flowers either solitary per branch, or arranged in a spike; perianth variously coloured, lasting at least one full day and fading slowly, not deliquescing; rootstock a corm or rhizome   -   -   -   -   -   -   -   -   -   7
7. Stamens included within a long perianth tube; the 3 style branches deeply forked and included or excluded   -   -   -   -   -   -   -   -   -   -   -   -   **7. Savannosiphon**
  – Stamens or at least the anthers exserted from the perianth tube; style simple, 3-branched or the branches again divided, the branches exserted   -   -   -   -   -   -   -   8
8. Flowers solitary on aerial axes; leaves junciform (terete, or oval in transverse section with 4 longitudinal grooves) and all inserted below the ground; perianth tube shorter than the tepals
  17. Romulea
  – Flowers 2–many, arranged in spikes or panicles; leaves usually with an expanded plane to plicate blade (sometimes leaves reduced to sheaths only), some at least inserted above ground level; perianth tube shorter or longer than the tepals   -   -   -   -   -   -   -   9
9. Style dividing at the mouth of the perianth tube into 3 long spreading branches; flowers actinomorphic   -   -   -   -   -   -   -   -   -   -   -   -   -   -   10
  – Style usually well exserted from the tube and either simple or dividing remotely from the tube into 3 branches, these sometimes again divided; flowers zygomorphic or sometimes actinomorphic   -   -   -   -   -   -   -   -   -   -   -   -   -   11
10. Perianth tube more than 20 mm long and straight; flowers scarlet, occasionally pink; tepals always more than 24 mm long   -   -   -   -   -   -   -   -   **16. Schizostylis**
  – Perianth tube 6–25 mm long, when more than 15 mm long then strongly curved apically; flowers white to lilac or pink; tepals always less than 20 mm long   -   -   -   **15. Hesperantha**
11. Style undivided to the apex; uppermost tepal much exceeding the others, longer than the perianth tube and strongly arcuate   -   -   -   -   -   -   -   **14. Zygotritonia**
  – Style divided into 3 branches, these sometimes again divided; uppermost tepal either as long as the others or exceeding them, shorter or longer than the tube, straight or arcuate   12
12. Corms campanulate with a flat base; stems somewhat compressed to angled or winged; inflorescences usually much branched and pseudopaniculate with 1–several sessile flowers on the ends of the main branches, or a cushion-like tuft borne at ground level   **8. Lapeirousia**
  – Corms globose to depressed globose with a rounded base; stems terete or winged; inflorescences spicate, the stems branched or not   -   -   -   -   -   -   -   -   -   -   13
13. Leaves pleated, at least lighly so   -   -   -   -   -   -   -   -   -   -   -   14
  – Leaves plane or terete, sometimes ribbed with the margins, midrib and other veins thickened (sometimes the leaves of the flowering stem lack blades and are entirely sheathing)   15
14. Stem largely underground; leaves lighly puberulent; flowers blue-violet and white
  13. Babiana
  – Stem aerial; leaves glabrous; flowers orange to reddish-brown   -   -   - **11. Crocosmia**
15. Flowers actinomorphic and often pendent or facing the ground but if upright or patent then the bracts scarious and translucent with brown streaking; perianth tube shorter than the tepals   -   -   -   -   -   -   -   -   -   -   -   -   -   -   -   16
  – Flowers zygomorphic, the stamens unilateral and arcuate to horizontal, perianth usually facing to the side but then bracts not scarious; perianth tube shorter or longer than the tepals   17
16. Flowers bell-shaped; stamens and style not exserted beyond the tepals; bracts scarious (or partly dry and membranous)   -   -   -   -   -   -   -   -   -   -   -   **9. Dierama**
  – Flowers hypocrateriform with the tepals spreading at right angles to the perianth tube; stamens and style well exserted; bracts coriaceous never scarious   -   -   -   - **11. Crocosmia**
17. Style branches each divided for about half their length, filiform throughout   **12. Anomatheca**
  – Style branches each simple or apically notched, filiform throughout or broader towards the apices   -   -   -   -   -   -   -   -   -   -   -   -   -   -   -   18
18. Lower tepals or at least the lower median tepals with an erect tooth-like callus in the midline; flowers shades of orange   -   -   -   -   -   -   -   -   -   **10. Tritonia**
  – Tepals without median calluses; flowers variously coloured, occasionally orange   19

19. Style branches usually broader towards the apices; foliage leaves variously developed, solitary to several or absent at flowering time, the blades plane, ribbed, lanceolate, linear or terete; perianth tube shorter or longer than the tepals; seeds broadly winged and thus discoid
19. Gladiolus
– Style branches filiform throughout; foliage leaves several, the blades plane and lanceolate; perianth tube longer than the tepals; seeds globose to angular **18. Radinosiphon**

1. ARISTEA Aiton

Aristea Aiton in Hort. Kew. ed. 1, **1**: 67 (1789). —Weimarck in Acta Univ. Lund. n.s., **36**,1 (1940). —Vincent in S. African J. Bot. **51**: 209–252 (1985).

Evergreen perennial herbs with short rhizomes, aerial parts persisting for more than one year. Leaves equitant, linear to lanceolate, distichous, mostly basal. Stems terete to compressed and 2-sided, often strongly winged, simple or branched, bearing reduced leaves or leafless. Inflorescences composed of binate rhipidia (umbellate flower clusters in 2 series, unless reduced to 1–2 flowers); binate rhipidia 1–many, terminal on the main and secondary axes, or axillary and then stalked or sessile; spathes (enclosing the binate rhipidia) herbaceous to membranous or scarious, entire or lacerate; floral bracts (within the spathes) membranous or scarious, entire or lacerate. Flowers frequently sessile, actinomorphic, blue, each lasting one morning only, the perianth twisting spirally on fading. Tepals basally connate for c. 1 mm or less, usually subequal, lanceolate to obovate, spreading horizontally. Stamens erect, filaments free, anthers oblong. Ovary ovoid to turbinate or cylindric, trigonous, often included in the bracts; style filiform, usually eccentric, dividing apically into 3 short stigmatic lobes. Capsules ovoid-ellipsoid to oblong-cylindric, usually 3-lobed (in some South African species 3-winged), included or exserted from the spathes, the remains of the perianth usually persisting on the capsules. Seeds few to many per locule, rounded to angular (laterally compressed in some South African species).

A genus of c. 50 species, best developed in South Africa, extending to Senegal and Ethiopia in the north, and with 6 species in Madagascar; most frequent in well watered highlands in grassland or rocky outcrops.

1. Flowering stem flattened and 2-winged, leafless except for 1(2) short subapical leaves or leafy bracts usually subtending the inflorescence; flower clusters 1–2 (rarely more) 7. *abyssinica*
– Flowering stem terete to weakly compressed or 2-angled (narrowly winged), bearing 2 or more leaves, these not subapical but mostly inserted below the middle; flower clusters usually more than 3 - - - - - - - - - - - - - - - - - - - 2
2. Flower clusters subtended by short leaves partly enclosing the spathes and bracts; capsules cylindric-trigonous, 12–18 mm long, on pedicels 6–10 mm long - - - 6. *ecklonii*
– Flower clusters subtended by inconspicuous, membranous to more or less dry spathes little different from the floral bracts; capsules ovoid-obovoid or oblong, 7–10 mm long, on pedicels more than 4 mm long - - - - - - - - - - - - - - 3
3. Spathes and bracts deeply lacerate; the lower flower clusters subtended by bract-like leaves longer than the spathes and bracts - - - - - - - - 5. *woodii*
– Spathes and bracts more or less entire, or if lacerate then the lower flower clusters not subtended by bract-like leaves - - - - - - - - - - - - 4
4. Stem simple, or with a few branches but then the axes straight; flower clusters spicately arranged (sessile except the terminal cluster); bracts shiny-white transparent 4. *nyikensis*
– Stem usually branched, usually repeatedly; flower clusters either all stalked, or a few below the terminal one sessile; spathes and bracts with dark solid keels and the edges transparent or shiny-white transparent - - - - - - - - - - - - - - - 5
5. Stem more or less dichotomously branched above; flower clusters seldom sessile - - - - - - - - - - - - - - - - 3. *polycephala*
– Stem unevenly branched with the central axis always dominant; usually at least 2 flower clusters sessile below the terminal one - - - - - - - - - - 6
6. Bracts without a raised midvein - - - - - - - - 1. *angolensis*
– Bracts with a raised midvein - - - - - - - - 2. *gerrardii*

1. **Aristea angolensis** Baker in Trans. Linn. Soc. London, ser. 2, Bot. **1**: 270 (1878); in F.T.A. **7**: 347 (1898). —Weimarck in Acta Univ. Lund. n.s., **36**,1: 18 (1940). —Hepper in F.W.T.A. ed. 2, **3**: 139 (1968). —Vincent in S. African J. Bot. **51**: 229 (1985). Type from Angola.
    Aristea majubensis Baker in J. Bot. **29**: 70 (1891). Type from South Africa (Natal).

Aristea cooperi Baker, Handb. Irid.: 143 (1892). Type from South Africa (Orange Free State).
Aristea nandiensis Baker in F.T.A. **7**: 347 (1898). Type from Kenya.
Aristea zombensis Baker in F.T.A. **7**: 346 (1898). —Rendle in J. Linn. Soc., Bot. **40**: 210 (1911).
—sensu Eyles in Trans. Roy. Soc. S. Africa **5**: 330 (1916) non Baker. Type: Malawi, Zomba Mt.,
xii.1898, *Whyte* s.n. (K, holotype).

[Subspecies admitted by Weimarck, l.c. 18 et seq. (1940) or Vincent, l.c. 229 et seq. (1985) are not
recognised for the Flora Zambesiaca area and their nomenclature is not listed.]

Plants 25–90 cm high. Leaves several, mostly basal and about half as long as the stem,
2.5–7(9) mm wide, linear to narrowly lanceolate, cauline leaves progressively shorter
above. Stem 2–6-branched (rarely simple), compressed, more or less elliptic in section,
2-angled but not winged, the branches fairly short, slender and erect. Flower clusters
4–14, terminal and axillary, 4–6-flowered; spathes 9–11 mm long, ovate, herbaceous with
scarious transparent margins, inner bracts 8–10 mm long, scarious and transparent
entirely or with herbaceous keels, the margins at first entire or lightly lacerate, becoming
increasingly so. Flower blue, more or less sessile. Tepals c. 12 × 4–5 mm, obovate.
Filaments 4 mm long, anthers c. 1.5 mm long. Style 6–7 mm long, exceeding the anthers,
apex 3-lobed. Capsules ovoid-obovoid, 5–7 mm long, sub-sessile or on pedicels up to 4
mm long.

Zambia. N: Mbala Distr., Nkali Dambo, 8.xi.1954, *Richards* 2176 (K; SRGH). W: Kitwe, 16.ii.1969,
Fanshawe 10521 (K; NDO; SRGH). **Zimbabwe**. N: Guruve (Chipolilo), Nyamunyeche, 19.xii.1978,
Nyariri 601 (K; SRGH). E: E of Stapleford road junction with Mutare-Nyanga road, 21.ii.1960, *Chase*
7279 (K; SRGH). C: Macheke, 1550 m, xii.1919, *Eyles* 1996 (K; SRGH). **Malawi**. N: Viphya Hills, 32
km SW of Mzuzu, 1600 m, 11.xi.1973, *Pawek* 7489 (K; MO; SRGH). S: Zomba Plateau, 1.xii.1966,
Banda 906 (K; MAL; SRGH).
    Also in South Africa (Natal, Orange Free State, Transvaal), Transkei, Angola, Tanzania, Kenya,
Ethiopia, eastern Nigeria and western Cameroon. Favouring wet sites and often in seeps, dambos or
shallow marshes, flowering in November to February.
    Distinguished from the other Flora Zambesiaca species of *Aristea* in the sessile flower clusters on
the main axis, the comparatively short and distinctively slender lateral branches, the short, partly
herbaceous, entire or slightly lacerate spathes, 9–11 mm long, and the weakly compressed stem, only
slightly angled but not winged.

2. **Aristea gerrardii** Weimarck in Acta Univ. Lund. n.s., **36**,1: 17 (1940). —Vincent in S. African J. Bot.
    51: 240 (1985). Type from South Africa (Natal).

Plants 15–100 cm high. Leaves several, mostly basal and about half as long as the stem,
6–12 mm wide, narrowly lanceolate to linear, cauline leaves progressively shorter above.
Stem flattened and 2-sided, branched repeatedly in the upper third, the main axis
predominant. Flower clusters sessile except those terminal on the axes, usually 4-
flowered; spathes and bracts scarious, the keels solid, transparent along the margins, 5–6
mm long. Flowers blue, more or less sessile. Tepals 7–12 mm long, obovate. Filaments 3–4
mm long, anthers c. 2.5 mm long. Style reaching to about the anther apices, 3-lobed.
Capsules oblong, 5–6, more or less sessile.

Mozambique. M: Matutuíne, between Zitundo and Manhoca, 29.xi.1979, *de Koning* 7706 (BR;
MO). Z: Pebane Distr., Pebane, viii.1950, *Munch* 261 (SRGH).
    Restricted to coastal southern and central Mozambique; also in South Africa (coastal Natal).
Flowering in August to December.
    Distinguished from the other Flora Zambesiaca species of *Aristea* in the repeatedly branched stem,
small laxly arranged rhipidia, and solid-keeled spathes with transparent entire margins.

3. **Aristea polycephala** Harms in Bot. Jahrb. Syst. **28**: 365 (1900). —Weimarck in Acta Univ. Lund.
    n.s., **36**,1: 15 (1940). Type from Tanzania.
    Aristea zombensis sensu Gomes e Sousa, Subsid. Estudo Fl. Niassa Port.: 56 (1935) non Baker.

Plants (40)50–100 cm tall. Leaves 6–8, basal and cauline, reaching to about the middle
of the stem, 3–8 mm wide, linear or falcate, the upper cauline leaves reduced in size and
partly to entirely sheathing. Stem repeatedly branched in the upper third, the central axis
predominant, weakly compressed, without wings. Flower clusters many, mostly stalked,
generally 2-flowered; spathes and bracts whitish and translucent with dark keels, 6–8 mm
long, the margins fimbriate. Flowers blue; tepals 8–10 mm long, obovate. Filaments c. 3
mm long, anthers 3 mm long. Style reaching the anther apices, 3-lobed. Capsules oblong,
c. 6 mm long.

Malawi. N: South Viphya near Nthungwa, 3.i.1987, *la Croix* 4267 (MAL; MO). **Mozambique**. N: Metónia, 1250 m, xii.1932, *Gomes e Sousa* 1108 (K).
Restricted in distribution to the Rift Valley mountains and escarpments to the NW and E of Lake Malawi, and SW Tanzania. Favouring marshy sites; flowering in December to February.

4. **Aristea nyikensis** Baker in Bull. Misc. Inform., Kew **1897**: 281 (1897); in F.T.A. **7**: 347 (1898). —Weimarck in Acta Univ. Lund. n.s., **36**,1: 22 (1940). Type: Malawi, Nyika Plateau, vii.1896, *Whyte* 230 (K, holotype).
    Aristea uhehensis Harms ex Engl. in Engler & Drude, Veg. der Erde 9, 2: 370 & fig. 26B (1908) nom. nud. Type from Tanzania.
    Aristea hockii De Wild. in Fedde, Repert. Spec. Nov. Regni Veg. **11**: 509 (1913). Type from Zaire.

Plants 20–30 cm high. Leaves 5–8, the lower 3–5 basal and narrowly lanceolate, the upper 2–3 cauline and nearly as large, linear, 4–6(8) mm wide, about two-thirds as long as the stem, fairly rigid, few-veined. Stem bearing 2–3 leaves, slightly compressed, simple or 1–2-branched. Flower clusters several per axis in a spicate arrangement, sessile except the terminal cluster, each (2)4-flowered; spathes and bracts scarious, pale and shiny, 7–8 mm long, ovate-lanceolate, the margins more or less entire. Flowers deep-blue, more or less sessile. Tepals c. 10 × 6 mm, obovate. Filaments 3.5 mm long, anthers 1.8 mm long. Style c. 5 mm long, 3-lobed apically. Capsules c. 7 mm long, ellipsoid-ovoid, on stalks up to 5 mm long.

Zambia. N: Chishinga Ranch, 11.i.1963, *Astle* 1936 (K; SRGH). W: Mwinilunga–Mutshatsha road, 12.xi.1962, *Richards* 17192 (K; SRGH). **Malawi**. N: Viphya Plateau, 45 km SW of Mzuzu, 23.ii.1974, *Pawek* 8318 (K; MO; SRGH). C: Dedza Distr., Chongoni Forest Reserve, 5.ii.1968, *Salubeni* 953 (MAL; SRGH).
Also in S Tanzania and Zaire (Shaba). Usually in seasonally moist localities, rock pools and streamsides; flowering in November to February.
Easily recognised by its scarious, silvery semi-transparent spathes and bracts generally lacking a dark keel, and the broad well-developed leaves on the flowering stem.

5. **Aristea woodii** N.E. Br. in Bull. Misc. Inform., Kew **1931**: 192 (1931). —Weimarck in Acta Univ. Lund. n.s., **36**,1: 28 (1940). —Vincent in S. African J. Bot. **51**: 225 (1985). TAB. 1 fig. B. Type from South Africa (Natal).
    Aristea torulosa var. *monostachya* Baker in F.C. **6**: 49 (1896). Type from South Africa (Natal).
    Aristea affinis N.E. Br. in Bull. Misc. Inform., Kew **1931**: 192 (1931). Type from South Africa (Transvaal).
    Aristea gracilis N.E. Br. in Bull. Misc. Inform., Kew **1931**: 193 (1931). Types from South Africa (Natal, Transvaal).

Plants 20–70 cm high. Leaves 6–10, mostly basal and about half as long as the stem, 3–5 mm wide, linear to falcate, fairly rigid; cauline leaves smaller, sometimes entirely sheathing. Stem lightly compressed, simple or with 1–2 short branches. Flower clusters 3–8, usually in a lax spicate arrangement, sessile except for the terminal cluster, (2)4-flowered, the lower flower clusters sometimes subtended by a prominent leafy bract 14–20 mm long; spathes 10–12 mm long, membranous with solid keels, the margins scarious and lacerate; bracts entirely scarious, the margins irregularly lacerate. Flowers deep-blue, more or less sessile. Tepals 10–14 × 5–7 mm, apparently obovate. Filaments 4.5–6 mm long, anthers 2 mm long. Style c. 6–7.5 mm long, 3-lobed. Capsule 6–10 mm long, ellipsoid-ovoid, on pedicels up to 2 mm long.

Zambia. W: Chingola, 9.xii.1965, *Fanshawe* 9457 (NDO; SRGH). **Zimbabwe**. E: Chimanimani Distr., Goma (Pork Pie), 25.xii.1948, *Chase* 1244 (K; SRGH). **Malawi**. N: Nyika Plateau, Dam II, 23.xi.1975, *Phillips* 342 (MO). **Mozambique**. MS: Chimanimani Mts., path from Skeleton Pass to Namadima, 1600 m, 30.xii.1959, *Goodier & Phipps* 348 (SRGH). M: Namaacha, Swaziland border, 600 m, ii.1931, *Gomes e Sousa* 411 (K).
Generally in wet habitats, either in dambos or along streams in submontane grassland. Extending into South Africa as far south as the eastern Cape Province.
Recognised by the large, deeply lacerate bracts and spathes subtended by fairly large bracts with membranous margins, and by the usually unbranched stems.

6. **Aristea ecklonii** Baker in J. Linn. Soc., Bot. **16**: 112 (1877); Handb. Irid.: 144 (1892). —Weimarck in Acta Univ. Lund. n.s., **36**,1: 58 (1940). —Hepper in F.W.T.A. ed. 2, **3**: 139 (1968). —Vincent in S. African J. Bot. **51**: 244 (1985). TAB. 1 fig. A. Type from South Africa (Cape Province).
    Aristea paniculata Pax in Bot. Jahrb. Syst. **15**: 151 (April, 1892). —Baker in F.T.A. **7**: 348

Tab. 1. A. —ARISTEA ECKLONII, whole plant (×⅔), from *Pope & Müller* 1686. B. —ARISTEA
WOODII, flowering spike (×⅔), from *Phillips* 346. C. —ARISTEA ABYSSINICA, whole plant
(×⅔), from *Gereau & Kayombo* 4437. Drawn by J.C. Manning.

(1898). —Fries, Wiss. Ergebn. Schwed. Rhod.-Kongo-Exped.: 235 (1916). Type: Mozambique, Namuli, Makuli country, 1887, *Last* s.n. (K, holotype).

Aristea lastii Baker, Handb. Irid.: 142 (Aug., 1892), nom. superfl. pro *A. paniculata.*

Aristea cyanea De Wild. in Pl. Bequaert. **1**: 52 (1921) nom. illegit. non *A. cyanea* Aiton. Type from Zaire.

Aristea stipitata R.C. Foster in Contrib. Gray Herb., No. 114: 50 (1936) as nom. nov. pro *A. cyanea* De Wild. nom. illegit.

Aristea maitlandii Hutch. in Hutchinson & Dalziel, F.W.T.A. ed. 1, **2**: 374 (1936), nom. illegit. (from Nigeria).

Plants 40–60 cm high. Leaves several, basal and cauline, the lower leaves longest and exceeding the stem in length, 5–9 mm wide, narrowly lanceolate, progressively reduced in size above, those subtending the axillary flower clusters 2–5 cm long and exceeding the spathes and bracts. Stem elliptic in section, usually winged, few- to several-branched. Flower clusters several to many, both terminal and axillary, stalked or more or less sessile in the axils, each usually (2)4-flowered; spathes scarious and rust-coloured with herbaceous keels, 13–16 mm long; floral bracts entirely scarious, c. 10 mm long, the margins entire. Flowers blue, on pedicels c. 5 mm long. Tepals 8–10 × c. 4 mm, obovate. Filaments c. 4 mm long, anthers c. 2 mm long. Style c. 4.5 mm long, reaching the middle of the anthers, 3-lobed. Capsules 12–18 mm long, cylindric-trigonous, on pedicels 6–10 mm long.

Zimbabwe. E: Chimanimani Distr., Glencoe Forest Reserve, 23.xi.1955, *Drummond* 4966 (K; SRGH). S: Mberengwa (Belingwe) Distr., Mt. Buhwa, 1585 m, 30.x.1973, *Gosden* 9 (MO; SRGH). **Mozambique**. MS: Chimanimani Mts., Musapa Gap, 6.x.1950, *Wild* 3522 (SRGH).

Extending from the eastern Cape Province (South Africa) to southern Zaire, Burundi, Tanzania and Uganda, and also in Cameroon. Typically along streams in riverine forest and evergreen forest margins, in shade; mostly flowering in October to December.

Easily recognised by the branched and leafy flowering stem and the long cylindric capsules up to 18 mm long.

7. **Aristea abyssinica** Pax in Engler, Hochgebirgsfl. Afrika: 173 (1892). TAB. **1** fig. C. Type from Ethiopia.

Aristea johnstoniana Rendle in Trans. Linn. Soc. London, ser. 2, Bot. **4**: 48 (1895). —Baker in F.T.A **7**: 346 (1898). Types: Malawi, Mt. Mulanje, *Whyte* 14 & 81 (BM, K).

Aristea tayloriana Rendle in Trans. Linn. Soc. London, ser. 2, Bot. **4**: 48 (1895). Type from Tanzania.

Aristea longifolia Baker in F.T.A. **7**: 576 (1898). Type from Tanzania.

Aristea bequaertii De Wild. in Fedde, Repert. Spec. Nov. Regni Veg. **11**: 509 (1913). Type from Zaire.

Aristea homblei De Wild. in Fedde, Repert. Spec. Nov. Regni Veg. **11**: 509 (1913). Type from Zaire.

Aristea cognata N.E. Br. ex Weimarck in Acta Univ. Lund. n.s., **36**, 1: 39 (1940). —Vincent in S. African J. Bot. **51**: 217 (1985). Type from South Africa (Transvaal).

Aristea alata subsp. *abyssinica* (Pax) Weimarck in Acta Univ. Lund. n.s., **36**, 1: 44 (1940). —Hepper in F.W.T.A. ed. 2, **3**: 139 (1968).

Aristea cognata subsp. *abyssinica* (Pax) Marais in Kew Bull. **42**: 932 (1987).

Plants 12–25(50) cm high. Leaves several, usually about half as long as the stem, 2–5(8) mm wide, linear to narrowly lanceolate, all basal except for one leaf (rarely 2) near the stem apex, this 3–5 cm long. Stem laterally compressed and broadly winged, 2–4 mm wide, unbranched or with a short subapical and often axillary branch subtended by a short subapical leaf. Flower clusters 1–2(3), terminal or subterminal, nearly sessile, 4-flowered; spathes 8–10 mm long, broadly lanceolate, scarious with herbaceous keels, margins entire in bud, becoming lacerate; floral bracts similar, entirely scarious. Flowers blue, more or less sessile. Tepals 10–12 × 6–7 mm, obovate. Filaments c. 4 mm long, anthers 1.5–2 mm long. Style c. 5 mm long, apex 3-fid. Capsules 7–9 mm long, ovoid-obovoid, more or less sessile or on short pedicels up to 4 mm long.

Zambia. N: Mbala Distr., Nkali Dambo, c. 1740 m, 21.ii.1967, *Richards* 22123 (K; MO). W: Kitwe, 18.i.1969, *Fanshawe* 10501 (NDO; SRGH). **Zimbabwe**. E: Nyanga, Rupango Mts., 1800 m, 3.ii.1952, *Chase* 4353 (MO; SRGH); Nyanga, World's View, Troutbeck, xii.1962, *Garley* 494 (K; SRGH). **Malawi**. N: 9 km N of Mzuzu to Lupaso, 1400 m, 7.iii.1976, *Pawek* 10885 (MAL; MO). C: Dedza, Chipazi Estate, 21.viii.1948, *Purse* s.n. (SRGH). S: Zomba Mt., 1520 m, 31.iii.1978, *Pawek* 14168 (K).

Also in South Africa (Transvaal, Natal, Cape Province), Zaire, East Africa, Ethiopia, and in highland Nigeria and Cameroon. Short plateau and submontane grassland, especially on thin rocky soils in highlands at elevations above 1500 m; mostly flowering in December to February.

Easily recognised by the broadly winged, leafless flowering stem and 1–3 terminal or nearly terminal flower clusters subtended by a short subapical leaf.

The taxonomy of the *A. abyssinica* complex requires careful investigation. Plants from South Africa and E Zimbabwe are sometimes treated as the separate species *A. cognata* but they appear to differ in no significant way from some collections from Malawi, Zambia and Uganda. Other collections from tropical Africa have slightly broader leaves and correspond with the type of *A. abyssinica* from Ethiopia. It seems best to adopt a broad circumscription for the complex and to recognise only one species here. The East African species *A. alata*, which is more robust and has broader more softly textured leaves, is undoubtedly closely related to *A. abyssinica*, as is the South African (Cape Province) species *A. anceps*, which has longer spathes and bracts up to 22 mm long, and capsules up to 12 mm long.

Specimens from Mt. Mulanje (Malawi), provisionally included here in *A. abyssinica*, (eg. *Chapman & Chapman* 8180 (K; MO) and *Chapman* H689 (SRGH)), have the general appearance of a robust form of *A. abyssinica*, but the stems, 35–45 cm high, are rounded rather than being flattened and 2-winged.

2. DIETES Salisb. ex Klatt

Dietes Salisb. [in Trans. Hort. Soc. **l**: 307 (1812) nom. nud.] ex Klatt in Linnaea **34**: 583 (1866) nom. conserv. vs. *Naron* Medik. —Goldblatt in Ann. Missouri Bot. Gard. **68**: 132–153 (1981).
Naron Medik. in Hist. & Comment. Acad. Elect. Sci. Theod.-Palat. **6**: 419 (1790).
Iris series *Dietes* (Salisb. ex Klatt) Baker in J. Linn. Soc., Bot. **l6**: 147 (1878).
Moraea subgen. *Dietes* (Salisb. ex Klatt) Baker, Handb. Irid.: 48 (1892); in F.C. **6**: ll (1896); in F.T.A. **7**: 342 (1898).

Medium to large evergreen perennial herbs with thick creeping rhizomes, aerial parts persisting for several years. Leaves several, distichous, ensiform, coriaceous, without a midrib. Stem usually erect, bearing leaves at lower nodes, with sheathing leaves at the upper nodes; branching irregularly in the upper half, or forming a divaricately branched panicle. Inflorescences composed of terminal rhipidia (umbellate flower clusters enclosed in paired opposed spathes); spathes sheathing, coriaceous, the outer spathe smaller than the inner. Flowers several per rhipidium, pedicellate, borne serially, actinomorphic, *Iris*-like; pedicels pubescent above; tepals free, all broadly unguiculate, the limbs spreading, outer tepal claws pubescent. Filaments usually free, occasionally united below, anthers linear, appressed to the style branches. Style dividing from near the base into 3 broad petaloid branches opposed to the outer tepals, each branch terminating in a pair of erect petaloid appendages (crests); stigmas transverse, abaxial, below the base of the crests. Capsules erect or pendent, cartilaginous, irregularly grooved or smooth, tardily dehiscent or indehiscent. Seeds irregularly angled, fairly large.

A genus of 6 species, 5 in southern and eastern tropical Africa and 1 on Lord Howe Island, Australasia.

Dietes grandiflora N.E. Br. with large white flowers (tepals 45–60 mm long), purple style branches and brown markings on the inner tepals, is often seen cultivated in gardens and street plantings. It is native to South Africa (E Cape Province and Natal). The flowers last for 2–3 days unlike the indigenous *D. iridioides*, which has similar but smaller flowers which last one day only.

Flowers white with yellow nectar guides on the outer tepals; style branches violet to blue; capsule erect, rough walled, usually furrowed and with a conspicuous beak   -   -   1. *iridioides*
Flower pale-yellow with orange-brown nectar guides; style branches cream to yellow (rarely blue-flushed); capsule pendent, smooth walled and without a beak   -   -   -   2. *flavida*

1. **Dietes iridioides** (L.) Sweet ex Klatt in T. Durand & Schinz, Consp. Fl. Afric. **5**: 156 (1895). —Goldblatt in Ann. Missouri Bot. Gard. **68**: 145 (1981). TAB. **2** fig. A. Type from South Africa.
   Moraea iridioides L., Mant.: 28 (1767). —Baker in F.T.A. **7**: 342 (1898). —Schinz & Junod in Mém. Herb. Boissier, No. 10: 30 (1900). —Rendle in J. Linn. Soc., Bot. **40**: 210 (1911). —Eyles in Trans. Roy. Soc. S. Africa **5**: 330 (1916). Type as above.
   Iris compressa L.f., Suppl.: 98 (1781). —Thunb., Diss. Irid. no. 12 (1782). Type from South Africa (Cape Province).
   Naron orientale Medik. in Hist. & Commentat. Acad. Elect. Sci. Theod. Palat. **6**: 419 (1790) nom. illegit. superfl. pro. *Moraea iridioides* L.

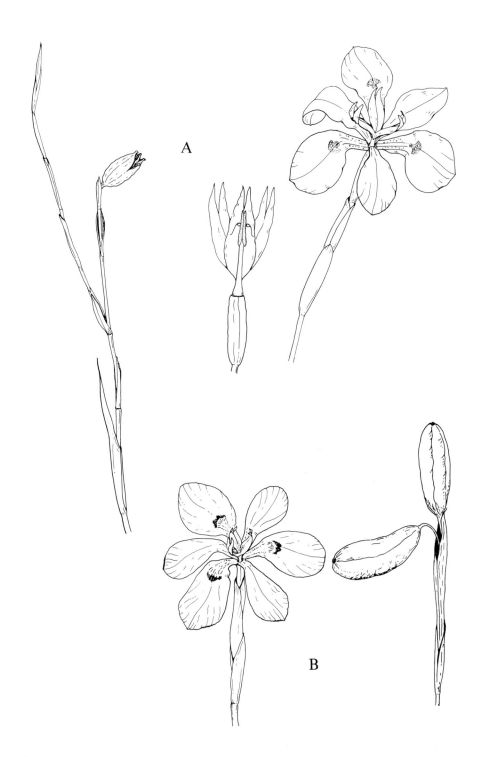

Tab. 2. A. —DIETES IRIDIOIDES, fruiting spike (×⅔), flower (×1), flower with tepals removed (×2), from *Pawek* 9677. B. —DIETES FLAVIDA, flower and capsules (×⅔), from *Bayliss* 8701. Drawn by Margo Branch.

Naron iridioides (L.) Moench, Meth. Pl.: 627 (1794), as *iridioideum"*. Type as for *Moraea iridioides* L.

Dietes iridifolia Salisb. in Trans. Hort. Soc., London **1**: 307 (1812) nom. inval. nom. nud. et superfl. pro *Moraea iridioides* L.

Moraea catenulata Lindl. in Bot. Reg.: tab. 1074 (1827). Type from Mauritius.

Dietes iridioides (L.) Sweet, Hort. Brit., ed. 2: 497 (1830). —Mogg in Macnae & Kalk, Nat. Hist. Inhaca Isl., Moçamb.: 143 (1958) *Dietes* nom. inval. nom. nud.

Dietes compressa (L.f.) Klatt in Linnaea **34**: 584 (1863). Type as for *Iris compressa* L.f.

Dietes catenulata (Lindl.) Sweet ex Klatt in Linnaea **34**: 585 (1863). Type as for *Moraea catenulata* Lindl.

Moraea iridioides var. *prolongata* Baker in F.C. **6**: 26 (1896). Type from South Africa (Natal).

Ferraria blanda Salisb., Prodr. Stirp.: 42 (1896) nom. illegit. superfl. pro. *Moraea iridioides* L.

Dietes vegeta sensu N.E. Br. in J. Linn. Soc., Bot. **48**: 36 (1928) quoad descr., non *Moraea vegeta* L.

Dietes prolongata (Baker) N.E. Br. in J. Linn. Soc., Bot. **48**: 37 (1928). —Brenan in Mem. N.Y. Bot. Gard. **9**, 1: 83 (1954).

Dietes prolongata var. *galpinii* N.E. Br. in J. Linn. Soc., Bot. **48**: 37 (1928). Type from South Africa (Transvaal).

Plants (15)30–60 cm high, often tufted. Leaves 25–40(60) × 6–15(25) mm, narrowly lanceolate. Stem irregularly branched, bearing short leaves below, and sheathing bract-like leaves above; sheathing leaves 25–30 mm long, often dry and brownish; old inflorescences sometimes producing branches with a terminal cluster of leaves, and ultimately rooting. Rhipidia spathes 35–50(55) mm long, the outer about half as long as the inner, obtuse to emarginate. Flowers white with yellow nectar guides on outer tepals, claws of outer tepals orange-dotted, style branches blue or white flushed with blue above; outer tepals 24–35 mm long, claw c. 16 mm long, limb 12–16 mm long, spreading to recurved; inner tepals 24–28 mm long, 9–12 mm wide, spreading to recurved. Filaments 5–9 mm long, free or united below, anthers 3–6 mm long. Ovary 8–15 mm long, lightly ridged; style 2–3 mm long, branches 7–9 mm long, 4–6 mm wide; crests c. 5 mm long. Capsule erect, 20–30 mm long, ovoid-cylindric, usually rostrate, surface fissured.

Zambia. W: Chingola, 4.x.1955, *Fanshawe* 2519 (K; NDO). **Zimbabwe**. C: Marondera (Marandellas), valley up to Wedza Mt., iii.1954, *Davies* 974 (LMA; SRGH). E: Nyanga, Holdenby, ix.1953, *Davies* 660 (MO; SRGH). S: Mberengwa (Belingwe), Mt. Buhwa, 10.xii.1953, *Wild* 4328 (K; MO; SRGH). **Malawi**. N: 8 km E of Mzuzu, 21.ix.1972, *Pawek* 5750 (K; MAL; MO; SRGH). C: Ntchisi Forest Reserve, 1645 m, 25.ii.1970, *Brummitt & Evans* 9398 (K; MAL; SRGH). S: Blantyre Distr., Bangwe Hill, 4 km E of Limbe, 1260 m, 23.xi.1977, *Brummitt, Seyani & Banda* 15153 (K; MAL). **Mozambique**. Z: Maganja da Costa, 8.ii.1966, *Torre & Correia* 14504 (LISC). MS: Mossurize, Espungabera, 11.x.1943, *Torre* 6154A (LISC; MO). M: Inhaca Island, 20.vii.1956, *Mogg* 31529 (K; SRGH).

Also in South Africa (Cape Province, Natal, Transvaal), eastern Zaire, Tanzania, Uganda, Kenya and Ethiopia. Usually in montane and coastal evergreen forests and forest margins; flowering irregularly, mostly in August to November.

Sometimes confused with the cultivated species *D. grandiflora* N.E. Br., from South Africa, which has larger flowers that last for two full days, and has brown markings on the inner tepals, and darker violet style branches.

2. **Dietes flavida** Oberm. in Fl. Pl. Africa 149: pl. 1488 (1967). —Goldblatt in Ann. Missouri Bot. Gard. **68**: 148 (1981). TAB. **2** fig. B. Type from South Africa (Natal).

Plants 50–70 cm high, forming clumps. Leaves 300–500 × 15–22 mm, narrowly lanceolate, glaucous. Stems branching irregularly, bearing long laminate leaves below, and reduced sheathing bract-like leaves above; stem bracts 30–50 mm long. Rhipidia spathes (40)45–50 mm long, the outer about half as long as the inner, obtuse to emarginate. Flowers pale-yellow, the outer tepals with brown nectar guides at the base of the limbs and spotted on the claws; outer tepals 30–40 mm long, claw c. 15 mm long, limb horizontal, 15–17 mm wide; inner tepals smaller, to 38 mm long. Filaments free, 4–6 mm long, broadened at the base, anthers 5–6 mm long. Ovary 10–14 mm long; style 2–3 mm long, branches c. 8 mm long and 3–4 mm wide; with crests 5–10 mm long, acute. Capsules pendent, 30–35 mm long, narrowly ovoid, smooth, dehiscing irregularly.

Mozambique. M: specimens have been seen from Namaacha, but no herbarium material is known.

Also in South Africa (Natal, Transvaal and possibly Cape Province) and Swaziland. Forest margins and lightly shaded areas.

Easily confused with *D. iridioides* but distinguished by the cream-coloured flowers and drooping, smooth-walled capsules.

3. FERRARIA Burm. ex Mill.

Ferraria Burm. ex Mill., Fig. Pl. Gard. Dict.: 187, pl. CCLXXX (1759). —de Vos in J. S. African Bot. **45**: 327–375 (1979).

Perennial herbs with depressed-globose corms, aerial parts dying back annually; corm tunics evanescent (lacking in herbarium collections). Leaves several to many, the lower 2–3 entirely stem-sheathing and membranous (cataphylls); foliage leaves (sometimes lacking at anthesis) equitant, lanceolate to linear, 1–several, with or without a midrib. Stems branched or simple, terete, sometimes sticky below the nodes. Inflorescence composed of several flower clusters (rhipidia); rhipidia terminal on the main and lateral axes, several-flowered; spathes herbaceous, the inner exceeding the outer. Flowers actinomorphic, often dull-coloured and speckled, usually unpleasantly scented. Tepals free, the outer 3 larger than the inner, unguiculate, the limbs usually with crisped margins. Filaments united into a column in the lower part, free and diverging above, anthers appressed to the style branches, the thecae parallel or divergent. Ovary included or excluded, style slender, concealed by the filament column, dividing into 3 flattened branches, these finely fringed above, the stigma lobe abaxial, at the base of the fringe. Capsules oblong-obovoid, sometimes rostrate.

A genus of 10 species, mostly of the west coast of South Africa (Cape Province) and SW Namibia. One species is widespread in central Africa.

Ferraria glutinosa (Baker) Rendle in Cat. Afr. Pl. Welw. **2**: 27 (1899). —Carter in Kew Bull. **17**: 317 (1963). —Sölch in Merxm., Prodr. Fl. SW. Afrika, fam. 155: 4 (1969). —Geerinck in Bull. Soc. Roy. Bot. Belg. **105**: 5 (1972). —de Vos in J. S. African Bot. **45**: 329 (1979). TAB. **3**. Type from Angola.
  Moraea glutinosa Baker in Trans. Linn. Soc. London, ser. 2, Bot. **1**: 271 (1878); Handb. Irid.: 55 (1892); in F.T.A. **7**: 342 (1898). —Klatt in T. Durand & Schinz, Consp. Fl. Afric. **5**: 150 (1895). Type as above.
  Moraea spithamaea Baker in Trans. Linn. Soc. London, ser. 2, Bot. **1**: 271 (1878); Handb. Irid.: 55 (1892); in F.T.A **7**: 341 (1898). Type from Angola.
  Moraea candelabrum Baker in Trans. Linn. Soc. London, ser. 2, Bot. **1**: 271 (1878); Handb. Irid.: 54 (1892); in F.T.A. **7**: 342 (1898). —Engler, Hochgebirgsfl. Trop. Afr.: 172 (1892). Type from Angola.
  Moraea andongensis Baker in Trans. Linn. Soc. London, ser. 2, Bot. **1**: 271 (1878). Type from Angola.
  Ferraria welwitschii Baker, Handb. Irid.: 74 (1892); in F.T.A. **7**: 344 (1898). —Rendle, Cat. Afr. Pl. Welw. **2**: 27 (1899). Type from Angola.
  Ferraria bechuanica Baker in F.T.A. **7**: 344 (1898). —N.E. Br. in Bull. Misc. Inform., Kew **1909**: 142 (1909). Type: Botswana, Ngamiland, Kalahari Desert near Mamunwe, *Lugard* 237 (K, holotype).
  Moraea randii Rendle in J. Bot. **36**: 144 (1898). Type: Zimbabwe, Bulawayo, i.1898, *Rand* 223 (BM, holotype; BR).
  Moraea aurantiaca Baker in F.T.A. **7**: 575 (1898). Type from Angola.
  Moraea kitambensis Baker in F.T.A. **7**: 575 (1898). Type from Angola.
  Ferraria candelabrum (Baker) Rendle in Cat. Afr. Pl. Welw. **2**: 27 (1899).
  Ferraria andongensis (Baker) Rendle in Cat. Afr. Pl. Welw. **2**: 27 (1899).
  Ferraria viscaria Schinz in Mem. Herb. Boissier, No. **10**: 77 (1900). Type from Namibia.
  Moraea malangensis Baker in Bull. Herb. Boissier, sér. 2, **1**: 862 (1901). Type from Angola.
  Ferraria randii (Rendle) Rendle in J. Bot. **43**: 54 (1905). —Eyles in Trans. Roy. Soc. S. Africa **5**: 330 (1916).
  Ferraria hirschbergii L. Bolus in S. African Gard. **22**: 57 & 59 (1932). Type from Zaire.

Plants (20)40–70(90) cm high. Corm brown, 10–40 mm in diameter, 2–4 internodes in length. Foliage leaves several (sometimes short or not developed at flowering time), usually about half as long as the stems, 4–8 mm wide, linear to narrowly lanceolate, decreasing in size above. Stem laxly and often repeatedly branched, sticky below the nodes. Rhipidia several, solitary on the branches, 2–6-flowered; outer spathes 15–25(30) mm long, obtuse to acute, inner spathes 30–45(50) mm long. Flowers brown, maroon, purple or yellow, usually spotted and mottled with contrasting colour, faintly scented; tepals lanceolate, the outer 28–35 mm long, the inner 25–28 mm long, the claws forming a wide cup c. 10 mm deep and 15 mm in diameter at the rim, limbs horizontal or recurved, the margins crisped. Filaments 10–13 mm long, united in the lower 8–10 mm, anthers 5 mm long, shrinking after anthesis to 2.5 mm long. Ovary 5–7 mm long; style c. 10 mm long, the branches 4 mm long. Capsules 12–20(25) mm long, globose-obovoid.

Tab. 3. FERRARIA GLUTINOSA, whole plant (×⅔), from *Smith* 3176 (floral details from *Goldblatt & Manning* 8805). Drawn by J.C. Manning.

Botswana. N: Groot Laagte (East), fossil river valley bed, 14–16.iii.1980, *P.A. Smith* 3176 (MO; SRGH). SW: 75 km N of Kang, *Wild* 5066 (SRGH). SE: 6 km S of Tshesebe (Tsessebe), 21.i.1960, *Leach & Noel* 298 (K; SRGH). **Zambia**. W: Kitwe, 3.xii.1959, *Fanshawe* 5314 (K; NDO). S: Mazabuka, Ridgeway Road, 2.xii.1931, *Trapnell* 541 (K). **Zimbabwe**. N: Mazowe Distr., Umvukwes, Ruorka Ranch, 17.xii.1952, *Wild* 3911 (MO; SRGH). W: Bulawayo, xii.1962, *Garley* 628 (K; SRGH). C: Gweru, c. 29 km SSE of Kwekwe, 23.xii.1965, *Biegel* 720 (K; SRGH). **Malawi**. N: c. 10 km N of Mpherembe, 1180 m, 7.ii.1987, *la Croix* 957 (MO).

Also in South Africa (N Cape Province), Zaire (Shaba), Angola and Namibia. In savanna and grassland, mostly in sandy soils, but also in stony ground.

4. MORAEA Mill.

Moraea Mill., Fig. Pl. Gard. Dict. **2**: 159, pl. CCXXIX (1758) (as "*Morea*") nom. conserv.
—Goldblatt in Ann. Missouri Bot. Gard. **64**: 243 (1977); in Ann. Kirstenbosch Bot. Gard. **14** (1986).

Perennial herbs with tunicate corms, aerial parts dying back annually. Leaves several to few, the lower 2–3 entirely sheathing and membranous (cataphylls); foliage leaves 1–several, inserted on the lower part of the stem, usually bifacial and channelled, sometimes terete; cauline leaves shorter or entirely sheathing and bract-like. Stem simple or branched (entirely subterranean in some South African species). Inflorescence(s) composed of rhipidia (umbellate flower clusters); rhipidia single per branch or occasionally crowded terminally, each enclosed in a pair of opposed leafy bracts (spathes) concealing the buds. Flowers *Iris*-like, usually blue or yellow with contrasting nectar guides on the outer tepals, radially symmetrical, usually pedicellate, borne serially. Tepals free (rarely united); the outer larger and strongly clawed with limbs spreading to reflexed; the inner often erect, or spreading. Filaments partly to completely united around the style (free in some South African species), anthers appressed to the style branches. Style filiform below, dividing into 3 flat usually petaloid branches, these diverging and usually forked apically into 2 crests (crests occasionally lacking), stigma transverse and abaxial below the crests. Capsules globose to cylindric. Seeds numerous, rounded, angular or discoid.

A genus of 120 species, 20 in tropical Africa and more than 102 in southern Africa.

Although all species are suspected of being poisonous to some extent, several are known to be toxic to stock, especially in the early stages of growth when sheep and cattle may become ill or die from eating them.

1. Foliage leaves terete, dry or absent at anthesis; axillary rhipidia sessile or nearly sessile 2
- Foliage leaves linear and channelled (or the margins inrolled on drying), appearing together with the flowers; axillary rhipidia, if present, on short to long branches - - - 3
2. Inner and outer tepals with nectar guides; style crests absent and style branches hardly wider than the anthers - - - - - - - - - - - 5. *thomsonii*
- Only the outer tepals with nectar guides; style crests 2–4 mm long and style branches wider than the anthers - - - - - - - - - - - 4. *stricta*
3. Flowers relatively small, the outer tepals 14–30(35) mm long; stem often branched 4
- Flowers relatively large, the outer tepals 35–90 mm long; stem simple with one terminal rhipidium (rarely with 1 branch) - - - - - - - - 8
4. Plants with 2–3 foliage leaves - - - - - - - - - 1. *carsonii*
- Plants with one foliage leaf - - - - - - - - - - 5
5. Foliage leaf inserted on the stem at or below ground level; sheathing (elaminate) leaves 5–8 cm long; flowers yellow - - - - - - - - - - 8. *inyangani*
- Foliage leaf inserted above the ground usually on the lower or middle third of the stem; sheathing leaves 2–4 cm long; flowers yellow or blue - - - - - 6
6. Flowers yellow; stem unbranched; spathes 4.5–6 cm long; outer tepals (2)2.5–3.5 cm long 16. *clavata*
- Flowers blue; stem usually branched; spathes 2–4.5(6) cm long; outer tepals 1.4–2.5(3) cm long - - - - - - - - - - - - - 7
7. Foliage leaf inserted on the lower third of the stem; branches crowded apically; spathes 4–6 cm long - - - - - - - - - - - - 2. *elliotii*
- Foliage leaf inserted on the upper third of the stem; branches laxly arranged; spathes 2–3.4 cm long - - - - - - - - - - - 3. *natalensis*
8. Bracts and spathes more or less dry at anthesis; flowers precocious, produced at the end of the dry season before or with the first rains (September to November) and before the foliage leaf emerges; the leaf shorter than the stem; cataphylls dark reddish-brown and prominent
17. *schimperi*

- Bracts and spathes green at anthesis; flowers produced during the wet season (December to May) after the foliage leaf has fully developed; the foliage leaf longer or shorter than the stem; cataphylls pale straw-coloured to brown, seldom prominent - - - - - 9
9. Foliage leaf shorter than the stem, sometimes almost entirely sheathing; sheathing (elaminate) leaves 1 or 2 - - - - - - - - - - - - - - - - 10
- Foliage leaf at least as long as the stem, usually much longer and usually bent and trailing; sheathing leaves (2)3–6 - - - - - - - - - - - - - - 11
10. Foliage leaf inserted on the stem well above the ground; sheathing leaves 1(2), not noticeably inflated - - - - - - - - - - - - - - - 15. *brevifolia*
- Foliage leaf usually inserted nearly at ground level; sheathing leaves 2, somewhat inflated
14. *tanzanica*
11. Flower colours in shades of blue to violet - - - - - - - - - 12
- Flowers yellow or whitish in colour - - - - - - - - - - 14
12. Outer tepals (5.7)6.5–8 cm long; anthers 12–15 mm long - - - - 11. *macrantha*
- Outer tepals 4–6.3 cm long; anthers 8–13 mm long - - - - - - 13
13. Inner tepals 4.8–6.5 cm; crests c. 1.5 cm long - - - - - - - 12. *textilis*
- Inner tepals 3.7–4.5 cm long; crests c. 1 cm long - - - - - - 13. *ventricosa*
14. Sheathing leaves not imbricate, 6–7(8) cm long; outer tepal limbs dark-spotted in the lower half
10. *bella*
- Sheathing leaves usually imbricate, generally 8–12 cm long; outer tepal limbs without dark-spotting - - - - - - - - - - - - - - - - 15
15. Outer tepals (5)6–10 cm long; anthers (10)11–14 mm long - - - - 9. *verdickii*
- Outer tepals 3.5–5.5(6) cm long; anthers 8–12 mm long - - - - - 16
16. Stems 80–150 cm high; leaf (10)12–16 mm wide - - - - - - 6. *spathulata*
- Stems (15)30–70 cm high; leaf (3)4–7 mm wide - - - - - - - 17
17. Inner tepals spathulate, 3.7–4.5 mm long; occurring in Zambia and N Malawi 13. *ventricosa*
- Inner tepals lanceolate, 3–3.5 mm long; occurring in E Zimbabwe (and South Africa)
7. *muddii*

Note. The collection of *Moraea polystachya* (L.f.) Ker Gawl. from "Botswana", *Pole Evans & Ehrens* 1931, is in fact from near Kuruman, in the northern Cape Province of South Africa, and was probably collected by Ehrens alone. However, the species may grow in Botswana but has not yet been recorded from there.

1. **Moraea carsonii** Baker in Bull. Misc. Inform., Kew **1894**: 391 (1894) as "*carsoni*"; in F.T.A. **7**: 341 (1898). —Goldblatt in Ann. Missouri Bot. Gard. **64**: 252–254 (1977). —Tredgold & Biegel, Rhod. Wild Fl.: 12, plate 7 (1979). —Goldblatt in Ann. Kirstenbosch Bot. Gard. **14**: 126 (1986). TAB. **4** fig. B. Type: Zambia, Fwambo, ix.1893, *Carson* s.n. (K, holotype).
   Moraea homblei De Wild. in Contrib. Fl. Katanga, Suppl. 4: 7 (1932). Type from Zaire (Shaba).

Plants 20–40 cm high. Corms 10–15 mm in diameter; tunics of fine to medium, dark-brown fibres. Cataphylls pale, often broken or fibrous above. Foliage leaves 2(3), longer than the stem but often trailing, 2–5 mm wide, linear, channelled; the lowermost basal or inserted near to well above the ground, the upper 1–2 leaves shorter than the basal leaf; sheathing leaves 2.5–6 cm long, green with dry brown attenuate apices. Stem erect, usually with 1–2(several) ascending branches. Rhipidia few to several, one per branch; spathes green and sometimes flushed with red, becoming dry and brown apically, the inner 3–4(5) cm long, outer c. 1 cm shorter than the inner. Flowers blue-violet with white to yellow nectar guides on the outer tepals; outer tepals 18–30 mm long, lanceolate, the limb slightly longer than the claw, spreading at c. 30° below horizontal; inner tepals 17–28 mm long, narrowly lanceolate to nearly linear, the limb also spreading. Filaments c. 7.5 mm long, free in the upper third, anthers 4.5–5.5 mm long. Ovary 3.5–7 mm long, exserted; style branches 8–10 mm long, crests 7–10 mm long. Capsule 7–9 mm long, globose.

Botswana. SW: Kobe Pan, Ghanzi – Kgalagadi Districts, 20.iii.1978, *Skarpe* S286 (K; PRE; UPS). **Zambia.** N: Mbala Distr., Chilongowelo Farm, 18.xii.1951, *Richards* 86 (K). W: Ndola, 2.i.1954, *Fanshawe* 612 (BR; K; NDO). C: Lusaka, 17.xiii.1956, *Angus* 1466 (K; LISC; SRGH). **Zimbabwe.** E: Nyanga Distr., Nyamaropa Forest Reserve, i.1966, *Wild* 7494 (BR; K; LISC; MO; PRE; SRGH). **Malawi.** N: c. 55 km SW of Mzuzu, 3.ii.1974, *Pawek* 8051 (K; MO; SRGH; WAG). C: Kongwe Mt., near Dowa, 18.ii.1959, *Robson & Steele* 1654 (BM; LISC; PRE; SRGH).
   Also in Zaire (Shaba), SW Tanzania and Namibia. Growing in open short grassland, often amongst rocks, or in light shade, sometimes in seasonally wet sites. Flowering in December to March.
   This species is similar in flower size, colour and shape to *Moraea elliotii* Baker and *Moraea natalensis* Baker, but distinguished by having 2–3 foliage leaves. Plants from the Nyanga Highlands in eastern Zimbabwe are especially robust and have unusually broad leaves and large flowers.

2. **Moraea elliotii** Baker, Handb. Irid.: 58 (1892); in F.C. **6**: 23 (1896). —Goldblatt in Ann. Missouri

Bot. Gard. **64**: 259 (1977); in Ann. Kirstenbosch Bot. Gard. **14**: 127 (1986). Type from South Africa (Transvaal).

Moraea macra Schltr. in J. Bot. **36**: 377 (1898). Type from South Africa (Cape Province).

Moraea violacea Baker in Bull. Herb. Boissier, sér. 2, **1**: 863 (1901). Type from South Africa (Natal).

Moraea juncifolia N.E. Br. in Trans. Roy. Soc. S. Africa **17**: 346 (1929). Type from South Africa (Transvaal).

Moraea stewartiae N.E. Br. in Trans. Roy. Soc. S. Africa **17**: 346 (1929). Type from Swaziland.

Plants 12–55 cm high. Corms 1–2 cm in diameter; tunics of dark-brown, medium to coarse fibres often extending upwards as a short neck. Cataphylls membranous, becoming dry and brown, ultimately fragmenting irregularly. Foliage leaf solitary, inserted near, to well above ground level, exceeding the inflorescence, linear and channelled to apparently terete with the margins tightly inrolled, an adaxial groove is usually present; sheathing leaves 2.5–4.5(6) cm long, green with dry brown attenuate apices. Stem more or less erect, simple or 1–3(8)-branched. Rhipidia single per ultimate branch; spathes green, becoming dry and brown above, the inner 4–5(6) cm long, the outer c. 1 cm shorter than the inner. Flowers blue-violet with yellow to orange nectar guides on the outer tepals; outer tepals 19–30 mm long, limb c. 18 × 9–12 mm, spreading; inner tepals 15–24 × 2–4 mm, narrowly lanceolate to nearly linear, limb spreading when fully open. Filaments 3.5–4 mm long, free in the upper half, anthers 5–6 mm long. Ovary 6–8 mm long, partly to fully exserted; style branches c. 8–10 mm long, crests 5–9 mm long. Capsule 9–12 × 4–5 mm, broadly obovoid.

Malawi: C: Dedza Distr., summit of Ciwawo (Ciwau) Hill, Chongoni Forest, 19.iii.1961, *Chapman* 1176 (SRGH).

Also in eastern South Africa, Lesotho and Swaziland. Usually occurring in moist grassland and rocky sites, especially in montane habitats, also in seeps and marshes. Flowering in February and March in the Flora Zambesiaca area.

3. **Moraea natalensis** Baker in Handb. Irid.: 56 (1892); in F.C. **6**: 20 (1896). —Goldblatt in Ann. Missouri Bot. Gard. **64**: 261 (1977); in Ann. Kirstenbosch Bot. Gard. **14**: 130 (1986). TAB. **4** fig. A. Type from South Africa (Natal).

Moraea erici-rosenii R.E. Fries, in Wiss. Ergebn. Schwed. Rhod.-Kongo-Exped. 1911–1912, **1**: 234 (1916). Type: Zambia, Kalambo, 1912, *Fries* 1345 (UPS, holotype).

Moraea parviflora N.E. Br. in Trans. Roy. Soc. S. Afr. **17**: 346 (1929). Type from South Africa (Transvaal).

Plants 15–45 cm high. Corms 10–15 mm in diameter; tunics of dark-brown to blackish fibres. Cataphylls membranous, usually unbroken. Foliage leaf solitary, inserted on the upper third of the stem, 5–20 cm long, shorter than or more often shortly exceeding the stem at anthesis, narrowly channelled, sometimes terete above; sheathing leaves seldom exceeding 2.5 cm in length, green at least below, dry and brown above. Stem erect or inclined, usually with crowded branches, the lowermost internode very long, spathes the internodes above the foliage leaf short. Rhipidia (1)3–several, crowded apically; spathes green, becoming dry above, apices attenuate, the inner 2.5–3.5 cm long, the outer c. 1 cm shorter than the inner, sometimes apically lacerate. Flowers grey-blue to violet with yellow nectar guides edged with dark-violet on the outer tepals; outer tepals 14–20 mm long, limb 7–14 × 6–10 mm, spreading c. 45° below the horizontal; inner tepals 9–15 mm long, narrowly lanceolate, the limb 4–6 mm wide, also spreading. Filaments 4–5 mm long, free in the upper half, anthers 4–5 mm long. Ovary c. 4 mm long, exserted; style branches 10–12 mm long including crests. Capsule 4.5–10 × 4–5 mm broadly ovoid.

Zambia. N: Kalambo Falls, c. 900 m, 3.xii.1964, *Richards* 19295 (K). W: Mwinilunga Distr., Zambezi Rapids, 10.xi.1966, *Richards* 17146 (K; SRGH). C: c. 8 km E of Lusaka, 30.xi.1955, *King* 229, (K). S: between Choma and Monze, 3.xii.1965, *van Rensburg* 3084 (BM; K; SRGH). **Zimbabwe**. W: Matobo Distr., farm Besna Kobila, xii.1953, *Miller* 1998 (K; LISC; MO; PRE; SRGH). C: 9 km SE of Gweru (Gwelo), 25.xi.1966, *Biegel* 1473 (MO; SRGH). S: near Masvingo (Fort Victoria), 14.xii.1968, *Plowes* 3154 (K; LISC; PRE; SRGH). **Malawi**. N: South Viphya, Lwanjati Peak, 11.i.1975, *Pawek* 8905 (K; MAL; MO; SRGH). C: near Mlanda (Tamanda) Mission, 8.i.1959, *Robson* 1093 (BM; K; LISC; PRE; SRGH). **Mozambique**. N: between Murrupula and Nivuraco R., 12.i.1961, *Carvalho* 426 (K; LMU). Z: Mugeba (Mujeba), c. 80 km from Namagoa on Pebane road, 23.xii.1947, *Faulkner* K152 (BR; K; PRE; S). MS: 12 km from Chimoio (Vila Pery), 24.xi.1965, *Torre & Correia* 13180 (LISC).

Also in eastern South Africa, (Transvaal and Natal) and in Zaire (Shaba). Occurring in seasonally wet localities (dambos), in exposed sites along rocky stream banks, and in seeps on rock outcrops. Flowering mainly in December and January; flowers opening in the late morning and fading after 4 pm.

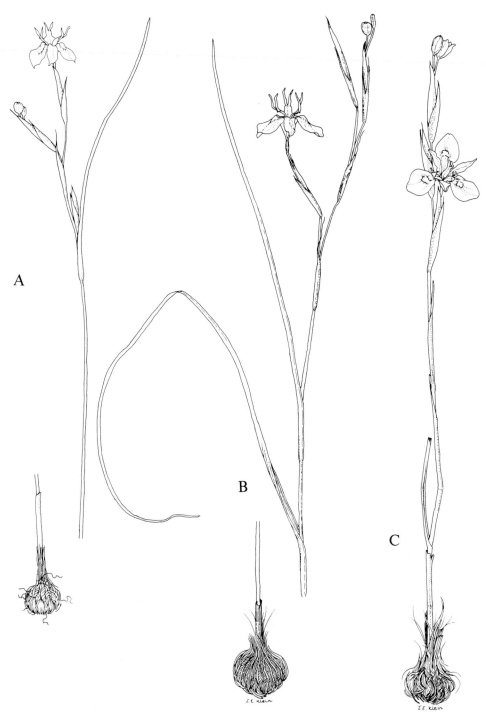

A

B

C

Tab. 4. A. —MORAEA NATALENSIS, whole plant (×½), from *Pawek* 8905. B. —MORAEA CARSONII, whole plant (×½), from *Pawek* 8051. C. —MORAEA STRICTA, habit (×½), from *Pawek* 10289. Drawn by J.E. Klein.

4. **Moraea stricta** Baker in Vierteljahrsschr. Naturf. Ges. Zürich **49**: 178 (1904). —Goldblatt in Ann. Kirstenbosch Bot. Gard. **14**: 134 (1986); in Fl. Pl. Africa **2**: plate 2069 (1993). TAB. **4** fig. C. Type from South Africa (Transvaal).

Moraea tellinii Chiov., Ann. Bot. Roma **9**: 138 (1911). Type from Ethiopia.

Moraea trita N.E. Br. in Trans. Roy. Soc. S. Afr. **17**: 347 (1929). —Letty in Wild Fl. Transvaal, tab. 36, fig. 1 (1962). Type from South Africa (Transvaal).

Moraea parva N.E. Br. in Trans. Roy. Soc. S. Afr. **17**: 347 (1929). Type from South Africa (Transvaal).

Moraea mossii N.E. Br. in Trans. Roy. Soc. S. Afr. **17**: 347 (1929). Type from South Africa (Transvaal).

Moraea curtisiae R.C. Foster in Contrib. Gray Herb., No. 127: 46 (1939). Type from Kenya.

Plants usually 15–25 cm high. Corms 1–3 cm in diameter; tunics of medium to coarse, dark-brown fibres, cormlets often present amongst the fibres. Foliage leaf solitary, usually absent or beginning to emerge at flowering time (dead leaf from previous season sometimes still attached to base of stem), eventually up to c. 60 cm long or more, c. 1.5 mm in diameter, terete, without an adaxial groove; sheathing leaves dry and brownish, 3–6 cm long, often apically lacerate. Stem erect, usually 3–6-branched; branches held close to the main stem. Rhipidia sessile or on short branches concealed by the subtending sheathing leaves; spathes dry and papery, rarely green near the base, the inner (2.5)3–4 cm long with apices membranous speckled darker brown and sometimes lacerated, the outer usually about two-thirds as long as the inner. Flowers pale-lilac to blue-violet with yellow-orange spotted nectar guides on the outer tepals; tepals unequal, the outer 19–24 mm long with an ascending claw narrow and slightly shorter than the limb, limb c. 11–14 × 5–8 mm, obovate to lanceolate, reflexed; the inner tepals 15–18 × 2–4 mm, linear-lanceolate, erect or ascending. Filaments 3–4 mm long, united in the lower third only, anthers 5–6 mm long. Style branches 7–8 mm long, diverging about 1.5 mm above the base; style crests narrow, 3–6 mm long. Capsules 8–11 mm long, obovoid.

Zambia. N: Lumi Marsh, Kawimbe, 6.ix.1956, *Richards* 6116 (K). E: Lundazi Distr., Nyika Plateau, Chowo Forest, 2200 m, 30.ix.1969, *Pawek* 2854 (K; MAL). **Zimbabwe**. E: Nyanga, Mare R., 3.ix.1952, *Wild* 3859 (K; MO; SRGH). **Malawi**. N: Nyika Plateau, 18.x.1975, *Pawek* 10289 (MO). S: Zomba Mt., 5.ix.1963, *Salubeni* 89 (MAL; SRGH). **Mozambique**. N: Maniamba, near Lichinga (Vila Cabral), 11.x.1942, *Mendonça* 776 (LISC). MS: Chimanimani Mts, between Skeleton Pass and Chimanimani Plateau, 27.ix.1966, *Grosvenor* 225 (K; LISC; SRGH).

Extending from the eastern Cape Province in South Africa, to northern Ethiopia. Usually in open, often stony submontane grassland and blooming among dry grasses at the end of the dry season; flowering in September to November in the Flora Zambesiaca region.

This species is recognised by the absence of a green leaf at flowering time, together with its sessile lateral inflorescences and small blue-violet to lilac flowers with unequal tepals, only the larger outer tepals having nectar guides. Easily confused with the vegetatively similar *Moraea thomsonii* Baker which differs in its pale-lilac to whitish flowers, its narrow style branches with crests less than 2 mm long or lacking, and its inner tepals also with nectar guides and usually as broad as the outer.

5. **Moraea thomsonii** Baker, Handb. Irid.: 57 (1892) as *"thomsoni"*; in F.T.A. **7**: 341 (1898). —Hemsley in Bot. Mag. **60**: t. 7976 (1904). —Brenan in Mem. N.Y. Bot. Gard. **9**, 1: 84 (1954). —Goldblatt in Ann. Missouri Bot. Gard. **64**: 262 (1977) excluding *M. stricta* and synonyms; in Ann. Kirstenbosch Bot. Gard. **14**: 137 (1986). Type from Tanzania.

Plants usually 15–25 cm high. Corms 10–30 mm in diameter; tunics of medium to coarse, dark-brown fibres, cormlets often present amongst the fibres. Foliage leaf solitary, usually absent or beginning to emerge at flowering time (dry leaf from previous season sometimes still attached to base of stem), eventually up to c. 60 cm long or more, 1.5–2 mm in diameter, terete, without an adaxial groove; sheathing leaves dry and brownish, 3–6 cm long, often apically lacerate. Stem erect, usually 3–6-branched; branches held close to the main stem. Rhipidia sessile or on short branches concealed by the subtending sheathing leaves; spathes dry and papery, rarely green near the base, the inner (2.5)3–4 cm long with apices membranous speckled darker brown and sometimes lacerated, the outer usually about two-thirds as long as the inner. Flowers pale-lilac to whitish with yellow-orange spotted nectar guides on the limbs of both whorls of tepals; tepals subequal, 19–24 mm long with an ascending claw narrow and slightly shorter than the limb, limbs c. 11–14 × 5–8 mm, obovate to lanceolate, spreading. Filaments 3–4 mm long, free only near the apex, anthers 5–6 mm long. Style branches c. 5 mm long, with crests less than 2 mm long or lacking. Capsule (8)9–11 mm long, obovoid.

Zimbabwe. E: Chimanimani (Melsetter) Distr., The Corner, 8.x.1950, *Chase* 2973 (K; MO; SRGH). **Malawi**. C: Nkhota Kota Distr., Chinthembwe (Chintembwe), 9.ix.1946, *Brass* 17579 (K; MO).

Also in South Africa (Transvaal) and SW Tanzania, in the highlands where they are apparently rare. Occurring in open submontane grassland, often in localities that remain moist throughout the dry season; flowering amongst dry grasses at the end of the dry season, in September to November. The flowers are said to open after sunset.

This species is recognised by the absence of a green leaf at flowering time, together with its sessile lateral inflorescences and small pale-lilac to white flowers with subequal tepals, both whorls of which have nectar guides, and also by the style crests being small or absent. Often confused with the more common, and vegetatively identical, *Moraea stricta* Baker, which has violet-lilac flowers with unequal tepals, only the outer whorl of which has nectar guides.

6. **Moraea spathulata** (L.f.) Klatt in T. Durand & Schinz, Consp. Fl. Afric. **5**: 152 (1895). —Goldblatt in Ann. Missouri Bot. Gard. **60**: 250 (1973); loc. cit. **64**: 250 (1977); in Ann. Kirstenbosch Bot. Gard. **14**: 206 (1986). —Plowes & Drummond, Wild Fl. Rhodesia: plate 43 (1976). TAB. **5** fig. A. Type from South Africa (Cape Province).

Iris spathulata L.f., Suppl. Pl.: 99 (1782). Type as above.
Iris spathacea Thunb., Diss. de Iride no. 23 (1782). Type from South Africa (Cape Province).
Moraea spathacea (Thunb.) Ker Gawl. in Bot. Mag. **28**: sub tab. 1103 (1808), non *Moraea spathacea* Thunberg (1787) (" *Bobartia indica* L.). —Baker in F.C. **6**: 14 (1896). Type as for *Iris spathacea* Thunb.
Moraea longispatha Klatt in Linnaea **34**: 560 (1866). Type from South Africa (Cape Province).
Moraea spathulata subsp. *transvaalensis* Goldblatt in Ann. Missouri Bot. Gard. **60**: 253 (1973). Type from South Africa (Transvaal).
Moraea spathulata subsp. *saxosa* Goldblatt in Ann. Missouri Bot. Gard. **60**: 254 (1973). Type from South Africa (Transvaal).
Moraea spathulata subsp. *autumnalis* Goldblatt in Ann. Missouri Bot. Gard. **60**: 254 (1973). Type from South Africa (Cape Province).

Plants large, 80–120 cm high, usually in small clumps or occasionally solitary. Corms 15–20 mm in diameter; tunics of brown, fine to coarse fibres. Cataphylls prominent, brown to pale, firm in texture, brittle, dry, entire or irregularly broken, or frayed at the apex. Foliage leaf solitary, much exceeding the stem in length and usually arching towards the ground, 12–16 mm wide, linear, flat or channelled; sheathing leaves 2–3, green or becoming dry and brown, 10–15 cm long, rarely imbricate. Stem simple (rarely 1-branched). Rhipidium solitary; spathes green or becoming dry and brown from the apex, attenuate, the inner 10–14 cm long, the outer about three-quarters as long as the inner. Flowers yellow with deep-yellow nectar guides on the outer tepals; outer tepals 35–50(60) mm long, the limb 20–35 mm long, spreading to reflexed; inner tepals 30–40 mm long, erect. Filaments 8–12 mm long, free in the upper half to one-third, anthers 8–12 mm long. Ovary 2–3 cm long, exserted from the spathes; style branches 12–18 mm long, the crests c. 10 mm long. Capsules 3.5–5.5 cm long, oblong-cylindric.

Zimbabwe. E: Mutare Distr., Banti North, 5.iii.1954, *Wild* 4522 (K; LISC; MO; PRE; SRGH). **Mozambique**. MS: Gorongoza Mt., iii.1972, *Tinley* 2436 (SRGH).

Also in southern Africa, extending as far south as George in the southern Cape Province. Mostly in well watered mountain areas in rocky grassland or at the edge of forest or bush, also coastal in South Africa; flowering in February to April in the Flora Zambesiaca region.

The large yellow flowers and a broad basal leaf usually much exceeding the stem and more or less flat above, together with the tendency to grow in small clumps distinguish this from other large yellow-flowered species of *Moraea* in tropical Africa. The cataphylls are usually dark-brown and often unbroken, or they may become irregularly fragmented from the apex.

7. **Moraea muddii** N.E. Br. in Trans. Roy. Soc. S. Afr. **27**: 346 (1929). —Goldblatt in Ann. Missouri Bot. Gard. **64**: 271 (1977); in Ann. Kirstenbosch Bot. Gard. **14**: 198 (1986). Type from South Africa (Transvaal).

Plants 15–50(70) cm high, usually c. 35 cm high. Corms 10–15 mm in diameter; tunics of pale straw-coloured fibres. Cataphylls brown, usually broken vertically into irregular strips. Leaf solitary, usually exceeding the stem in length, 3–6 mm wide, linear, channelled; sheathing leaves 2–3(4), green, with a dry brown apex, 8–12 cm long. Stem erect, simple. Rhipidium single, terminal; spathes green with a dry brown apex, the inner 7–9(12) cm long, the outer slightly shorter than the inner. Flowers cream to yellow with deeper yellow nectar guides on the outer tepals; outer tepals 35–50 mm long, lanceolate, claw 12–20 mm long, limb slightly reflexed, wide; inner tepals c. 35 × 7 mm long, lanceolate, erect. Filaments c. 10 mm long, united in the lower two-thirds, anthers 8–9 mm long. Ovary c. 15 mm long, oblong, exserted from the spathes; style branches 11–13 mm long, crests c. 10 mm long. Capsules c. 2 cm long, cylindric.

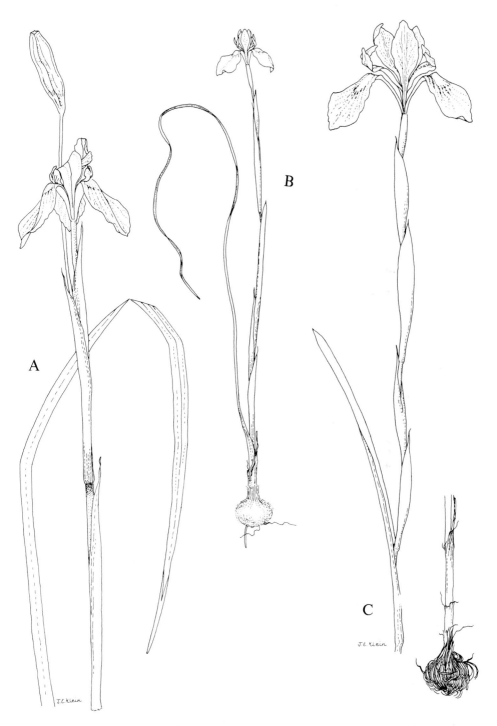

Tab. 5. A. —MORAEA SPATHULATA, flower shoot and leaf (×½), from *Wild* 4522. B. —MORAEA INYANGANI, habit (×½), from *Drummond & Robson* 5830. C. —MORAEA TANZANICA, whole plant (×½), from *Phillips* 1071A. Drawn by J.E. Klein.

Zimbabwe. E: Chimanimani Distr., Bundi R. banks, 29.viii.1964, *Whellan* 2153 (SRGH).
Mozambique. MS: Near summit of Chimanimani Mts., 25.ix.1966, *Grosvenor* 195 (LISC; SRGH; UPS; WAG).
Also in eastern South Africa as far south as the Amatola Mts. in E Cape Province, but most common in the Transvaal. In open montane grassland, in peaty or stony ground, usually at elevations above 1800 m; flowering in mid-September to December, rarely later.

8. **Moraea inyangani** Goldblatt in Ann. Missouri Bot. Gard. **64**: 272 (1977). TAB. **5** fig. B. Type: Zimbabwe, Mt. Inyangani, c. 2400 m, x.1961, *Wild* 5519 (SRGH, holotype; K; LISC; M; PRE).

Plants 15–30 cm high. Corms to 10 mm in diameter; tunics of fine pale straw-coloured fibres. Cataphylls brown, usually irregularly broken. Foliage leaf solitary, longer than the stem, c. 3 mm wide, linear, channelled; sheathing leaves 3(4), imbricate, green, 5–8 cm long. Stem erect, unbranched. Rhipidium single, terminal; spathes green with a dry apex, the inner 5–8 cm long, the outer slightly shorter than the inner. Flowers pale-yellow with deeper yellow nectar guides on the outer tepals; outer tepals to 25 mm long, lanceolate, claw about 10 mm long, limb 15 mm long, slightly reflexed; inner tepals 15–20 mm long, lanceolate, erect. Filaments 4 mm long, united in the lower two-thirds, anthers c. 6 mm long. Ovary c. 14 mm long, oblong, exserted; style branches 8–9 mm long, crests about 4–8 mm long. Capsules unknown.

Zimbabwe. E: Nyanga, summit ridge of Mt. Inyangani, ix.1958, *Drummond & Robson* 5830 (K; PRE; SRGH).
A rare local endemic of eastern Zimbabwe, occurring at higher elevations on Mt. Inyangani in the Nyanga Highlands, in seeps, marshy localities and along streams above 2400 m. Flowering in September, October and in April.
This species is distinctive in its low stature and relatively small flowers with outer tepals up to 25 mm long.

9. **Moraea verdickii** De Wild. in Ann. Mus. Congo, Bot., Sér. 4, **1**: 17 (1902). —Goldblatt in Ann. Missouri Bot. Gard. **64**: 277 (1977). Type from Zaire (Shaba).

Plants 45–75 cm high. Corms c. 20 mm in diameter; tunics of coarse pale or dark-brown fibres. Cataphylls pale to dark-brown, dry, broken and becoming fibrous above. Foliage leaf solitary, shortly exceeding the stem, 8–12 mm wide, linear, channelled or flat; sheathing leaves 2–3(4) usually widely spaced, green with a dry brown apex, 9–11 cm long. Stem erect, simple. Rhipidium single, terminal; spathes green with a dry brown apex, the inner 9–15 cm long, the outer c. two-thirds as long as the inner. Flowers yellow with deeper yellow nectar guides on the outer tepals; outer tepals (50)60–75(100) mm long, lanceolate, the limb 30–50 mm long; inner tepals 45–70 mm long, lanceolate, erect. Filaments 11–16 mm long, free in the upper third, anthers 10–14 mm long. Ovary 15–25 mm long, usually exserted; style branches 17–25 mm long, crests 10–17 mm long. Capsules unknown.

Zambia. N: Mbala Distr., Itembwe Gorge, c. 1500 m, 3.i.1960, *Richards* 12055 (EA; K; MO; SRGH). W: Mwinilunga Distr., Matonchi Farm, 2.i.1938, *Milne-Redhead* 3930 (BM; BR; K; LISC; PRE). **Malawi**. C: Dedza Distr., Chongoni Forest, 7.ii.1969, *Salubeni* 1253 (K; MAL; SRGH). S: Kirk Range, Ntcheu–Neno road, c. 1700 m, 31.i.1959, *Robson* 1404A (K; LISC; SRGH). **Mozambique**. N: Lichinga (Vila Cabral), ii.1934, *Torre* 39 (WAG). T: 8 km from Mlangeni on Ntcheu–Dedza road, 18.ii.1970, *Brummitt* 8609 (K; SRGH). Z: near Gurué (Vila Quanquiero), 1943, *Torre* 5078 (LISC).
Also in eastern Angola, Zaire (Shaba) and SW Tanzania. In rocky grassland, forest margins and light woodland, usually above 1200 m, also in seasonally wet localities; flowering mainly in December to February.

10. **Moraea bella** Harms in Bot. Jahrb. Syst. **28**: 364 (1901). —Goldblatt in Ann. Missouri Bot. Gard. **64**: 274 (1977). TAB. **6** fig. A. Type from Tanzania.

Plants (30)40–70 cm high. Corms c. 15 mm in diameter; tunics of pale medium to coarse fibres. Cataphylls light-brown, irregularly broken. Foliage leaf solitary, much exceeding the stem, 2.5–6 mm wide, linear, channelled; sheathing leaves 2–4, green, 6–7 mm long. Stem erect, simple. Rhipidium single, terminal; spathes green with a dry brown apex, the inner 6–10 cm long, the outer c. two-thirds as long as the inner. Flowers pale-yellow, the outer tepals with deeper yellow nectar guides and the limbs conspicuously darkly veined and spotted in the lower half; outer tepals 45–55(65) mm long, lanceolate, the limb 25–40 mm long, often slightly exceeding the claw; inner tepals 35–50(60) mm long, lanceolate, erect. Filaments 9–14 mm long, free in the upper third, anthers 8–10 mm long. Ovary

15–20 mm long, usually exserted; style branches 13–16 mm long, crests 8–15 mm long. Capsules 20–30 mm long, oblong.

Zambia. N: Mbala Distr., Kawimbe Road, c. 1450 m, 1.ix.1966, *Richards* 21407 (K; MO). C: Chakwenga Headwaters, E of Lusaka, 27.iii.1966, *Robinson* 6473 (K; SRGH). E: Chipata (Fort Jameson), 1.vi.1958, *Fanshawe* 4501 (EA; K; NDO; SRGH). **Malawi**. N: Mzimba Distr., Malembo, 20.iv.1974, *Pawek* 8460 (K; MO; SRGH). C: Kasungu Game Reserve, 21.vi.1970, *Brummitt* 11609 (K). **Mozambique**. N: Lichinga (Vila Cabral), 17.v.1948, *Pedro & Pedrógão* 3629 (EA; LMU).
Also in SW Tanzania and Zaire (Shaba). Restricted to wet habitats, marshes, dambos and streambanks; flowering late in the season, mostly in April to June.

11. **Moraea macrantha** Baker in F.T.A. **7**: 340 (1898). —Goldblatt in Ann. Missouri Bot. Gard. **64**: 279 (1977). Type: Malawi, without precise locality, without date, *Whyte* s.n. (K, holotype).

Plants 50–70 cm high. Corms c. 15 mm in diameter; tunics of pale fine to medium-textured fibres. Cataphylls brown, irregularly broken to more or less fibrous. Foliage leaf much exceeding the stem, 5–7 mm wide, linear, channelled; sheathing leaves 4–5 imbricate, 6–10 cm long, green. Stem erect, simple. Rhipidium single, terminal; spathes green with a dry brown apex, the inner 9–13 cm long, the outer ± two-thirds as long as the inner. Flowers blue to violet with white to pale-yellow nectar guides on the outer tepals; outer tepals (57)65–80 mm long, lanceolate, the limb 30–45 mm long, equal to or longer than the claw; inner tepals 55–75 mm long, lanceolate, erect. Filaments 13–17 mm long, free in the upper half to one third, anthers 12–15 mm long. Ovary c. 20 mm long, often included in the spathes; style branches c. 25 mm long, crests c. 20 mm long. Capsules c. 3 cm long, narrowly ovoid.

Zambia. E: Nyika Plateau, 2190 m, 18.ii.1961, *Richards* 14388 (K). **Malawi**. N: Viphya Plateau, 25.iv.1967, *Salubeni* 660 (K; LISC; MAL; PRE; SRGH).
Restricted to northern Malawi and adjacent Zambia; also in SW Tanzania and Zaire (eastern Shaba). In high altitude woodland and montane grassland, generally above 1800 m; flowering late in the season, usually in March to June.

12. **Moraea textilis** Baker in Trans. Linn. Soc. London, ser. 2, Bot. **1**: 270 (1878). —Goldblatt in Ann. Missouri Bot. Gard. **64**: 281 (1977). Type from Angola.
   Moraea mechowii Pax in Bot. Jahrb. Syst. **15**: 151 (1893). —Baker in F.T.A. **7**: 339 (1898). Type from Angola.

Plants 40–80 cm high. Corms c. 20 mm in diameter; tunics of coarse grey wiry fibres. Cataphylls pale with dark veins, irregularly broken. Foliage leaf solitary, exceeding the stem, 6–8 mm wide, linear, channelled; sheathing leaves (3)5–7, green at anthesis, 7–8 mm long. Stem erect, simple. Rhipidium single, terminal; spathes green with a dry brown apex, the inner 8–15 cm long, the outer c. three quarters as long as the inner. Flowers blue-purple, or yellow with deeper yellow nectar guides on the outer tepals; outer tepals 47–63 mm long, lanceolate, the claw often slightly exceeding the limb; inner tepals 48–65 mm long, lanceolate, erect. Filaments 15–22 mm long, free in the upper quarter, anthers 10–13 mm long. Ovary c. 2 cm long, usually exserted; style branches 1.5–2.5 cm long, crests 10–15 mm long. Capsules c. 3 cm long, narrowly ovoid.

Zambia. W: Mwinilunga Distr., Plains, iv–vi.1929, *Marks* 42 (K).
Also in Angola where it is common. In rocky grassland, usually above 1200 m, also recorded in seasonally wet sites; flowering in November to March.

13. **Moraea ventricosa** Baker in Bull. Misc. Inform., Kew **1895**: 73 (1895); in F.T.A. **7**: 340 (1898). —Goldblatt in Ann. Missouri Bot. Gard. **64**: 280 (1977). Type: Zambia, Fwambo, in 1894, *Carson* 37 (K, holotype).
   Moraea bequaertii De Wild. in Fedde, Repert. Spec. Nov. Regni Veg. **11**: 540 (1913). Type from Zaire.

Plants 30–55 cm high. Corms c. 15 mm in diameter; tunics of fine to medium pale straw-coloured fibres. Cataphylls pale to dark-brown, broken and becoming fibrous. Foliage leaf solitary, exceeding the stem, 3–7 mm wide, linear, channelled; sheathing leaves 3–6, imbricate, green, 5–8 cm long. Stem erect, unbranched. Rhipidium single, terminal; spathes green with dry apices, the inner 8–12(13.5) cm long, the outer c. two-thirds as long as the inner. Flowers usually blue-purple (or white to pale-yellow) with white to yellow nectar guides on the outer tepals; outer tepals 40–55 mm long, lanceolate, claw usually slightly longer than the limb; inner tepals 37–44 mm long, strongly

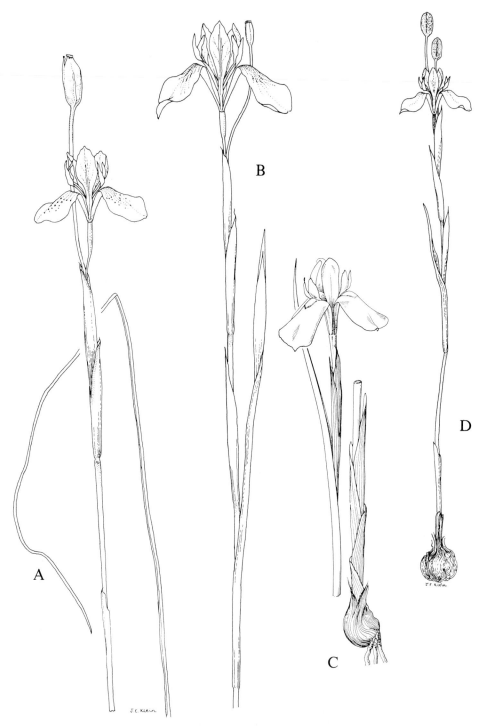

Tab. 6. A. —MORAEA BELLA, flower shoot and leaf (×½), from *Pawek* 8460. B. —MORAEA
BREVIFOLIA, flower shoot (×½), from *Simon & Williamson* 1483. C. —MORAEA SCHIMPERI,
upper and basal parts of flower shoot (×½), from *Goldblatt* 4593. D. —MORAEA CLAVATA,
habit (×½), from *Welwitsch* 1545. Drawn by J.E. Klein.

spathulate, erect. Filaments 12–15 mm long, united in the lower two-thirds, anthers 8–10(11) mm long. Ovary c. 20 mm long, oblong, usually included; style branches 14–17 mm long, crests c. 10 mm long. Capsules c. 3 cm long, narrowly ovoid.

Zambia. N: Kawimbe, 1800 m, 19.ii.1964, *Richards* 19038 (K; MO). W: Ndola, 14.iii.1954, *Fanshawe* 969 (BR; EA; K; NDO; SRGH).

Also in Tanzania, Burundi and Zaire (Shaba). In light deciduous woodland and grassland, often in seasonally wet sites such as dambo margins; flowering mostly in March to May.

14. **Moraea tanzanica** Goldblatt in Ann. Missouri Bot. Gard. **64**: 283 (1977). TAB. **5** fig. C. Type from Tanzania.

Plants 20–35 cm high. Corms c. 15 mm in diameter; tunics of pale medium-textured fibres. Cataphylls pale straw-coloured, irregularly broken. Foliage leaf solitary, 15–20 cm long, shorter or slightly longer than the stem, 3–6 mm wide, linear, channelled; sheathing leaves 2, green, 5–8 cm long. Stem erect, simple. Rhipidium single, terminal; spathes green with dry brown apices, the inner 10–12 cm long, the outer c. three quarters as long as the inner. Flowers pale-yellow with deeper yellow nectar guides on the outer tepals; outer tepals 45–55 mm long, lanceolate, the limb slightly exceeding the claw; inner tepals 33–45 mm long, lanceolate, erect. Filaments 12–15 mm long, free in the upper half, anthers c. 10 mm long. Ovary 17–22 mm long, usually exserted; style branches 17–20 mm long, crests c. 13 mm long. Capsules 2.5–3 cm long, narrowly ovoid-oblong.

Malawi. N: Nyika Plateau, Katumbi, Juniper Forest turnoff, 31.i.1976, *Phillips* 1071A (K; MO; SRGH).

Also in SW Tanzania. In open mountane grassland above 2200 m; flowering in January to March. This species is distinctive in its relatively short stature, its 2–3 imbricate sheathing leaves and its short foliage leaf seldom exceeding the stem at anthesis.

15. **Moraea brevifolia** Goldblatt in Ann. Missouri Bot. Gard. **64**: 285 (1977). TAB. **6** fig. B. Type: Zambia, Mporokoso Distr., Lumangwe Falls, 21.xii.1967, *Simon & Williamson* 1483 (K, holotype; LISC; SRGH).

Plants 20–50 cm high. Corms 10–15 mm in diameter; tunics of fine brown reticulate fibres. Cataphylls brown, dry, lacerated above. Foliage leaf solitary, inserted shortly above the ground and reaching to about the middle of the stem, 2–6 mm wide, linear, channelled; sheathing leaves 1(2), green, 6–9 cm long. Stem erect, simple. Rhipidium single, terminal; spathes green with dry brown apices, the inner 7.5–12 cm long, the outer c. two-thirds as long as the inner. Flowers pale-yellow with deeper yellow nectar guides on the outer tepals; outer tepals 50–70 mm long, lanceolate, the limb 30–40 mm long, slightly exceeding the claw; inner tepals 40–50 mm long, lanceolate, erect. Filaments 10–12 mm long, free in the upper third, anthers 9–16 mm long. Ovary 10–13 mm long, usually exserted; style branches 12–25 mm long, crests c. 15 mm long. Capsules unknown.

Zambia. N: Kambole–Mbala road, 7.xii.1959, *Richards* 11901 (K). W: Mwinilunga Distr., 2.i.1938, *Milne-Redhead* 3930 (BM; BR; LISC).

Restricted to northern and northwestern Zambia. Occurring in marshy habitats; flowering in December and January.

16. **Moraea clavata** R.C. Foster in Contrib. Gray Herb., No. 114: 49 (1936) as nom. nov. pro *Moraea gracilis* Baker nom. illegit. —Goldblatt in Ann. Missouri Bot. Gard. **64**: 287 (1977). TAB. **6** fig. D. Type from Angola.

Moraea gracilis Baker in Trans. Linn. Soc. London, ser. 2, Bot. **1**: 271 (1878); in F.T.A. **7**: 338 (1898) nom. illegit. non *Moraea gracilis* (Licht. ex Roem. & Schult.) Diels (1833). Type as above.

Plants 15–35 cm high. Corms 10–15 mm in diameter; tunics of fine light-brown reticulate fibres. Cataphylls pale and membranous, or green above, usually unbroken. Foliage leaf solitary, 5–12(20) cm long, inserted in the middle or upper part of the stem and rarely exceeding it in length, 3–5 mm wide, channelled; sheathing leaf 1, 3–4 cm long. Stem erect, simple. Rhipidium single, terminal; spathes green with dry brown apices, the inner 4.5–7 cm long, the outer c. two-thirds as long as the inner. Flowers pale-yellow with deeper yellow nectar guides on the outer tepals; outer tepals 20–30(35) mm long, lanceolate, the claw often slightly exceeding the limb; inner tepals 13–20 mm long, lanceolate, erect. Filaments 5–6 mm long, free in the upper half, anthers 4–5 mm long. Ovary 5–8 mm long, usually exserted; style branches c. 10 mm long, crests c. 10 mm long. Capsules 7–10 mm long, globose to ovoid.

Zambia. W: Mwinilunga Distr., dambo NE of Dobeka bridge, 11.xii.1937, *Milne-Redhead* 3620 (K). C: Serenje, 27.ix.1961, *Fanshawe* 6731 (K; NDO; SRGH).

Also in Angola. In moist habitats, stream banks, dambo margins and seeps; flowering in October to December, occasionally in January.

17. **Moraea schimperi** (Hochst.) Pic. Serm. in Webbia **7**: 349 (1950). —Brenan in Mem. N.Y. Bot. Gard. **9**, 1: 83 (1954). —Goldblatt in Ann. Missouri Bot. Gard. **64**: 268–270 (1977); in Ann. Kirstenbosch Bot. Gard. **14**: 216 (1986). TAB. **6** fig. C. Type from Ethiopia.

    Hymenostigma schimperi Hochst. in Flora **27**: 24 (1844). Type as above.

    Hymenostigma tridentatum Hochst. in Flora **27**: 25 (1844). Type from Ethiopia.

    Vieusseuxia schimperi (Hochst.) A. Rich., Tent. Fl. Abyss. **2**: 305 (1850).

    Vieusseuxia tridentatum (Hochst.) A. Rich., Tent. Fl. Abyss. **2**: 305 (1850).

    Xiphion diversifolium Steud. ex Klatt in Linnaea **34**: 572 (1866) nom. illegit. superfl. pro *Hymenostigma schimperi* Hochst.

    Moraea diversifolia (Steud. ex Klatt) Baker in J. Linn. Soc., Bot. **16**: 130 (1877); in F.T.A. **7**: 339 (1898). nom. illegit.

    Moraea welwitschii Baker in Trans. Linn. Soc. London, ser. 2, Bot. **1**: 270 (1877); in F.T.A. **7**: 339 (1898). Type from Angola.

    Moraea zambeziaca Baker in F.T.A. **7**: 339 (1898). —Fries, Wiss. Ergebn. Schwed. Rhod.-Kongo-Exped.: 234 (1916). Type: Malawi, Manganja Hills, ix–xi.1861, *Meller* s.n. (K, lectotype here designated, "entrance of Bangwe (Bangue) Pass"; K, isolectotype).

    Moraea hockii De Wild. in Fedde, Repert. Spec. Nov. Regni Veg. **11**: 540 (1913). Type from Zaire (Shaba).

    Moraea mechowii sensu Suessenguth & Merxmüller, Contrib. Fl. Marandellas Distr.: 76 (1951) non Pax (*M. textilis* Baker, from Angola).

Plants 20–50 cm high, sometimes caespitose. Corms 15–20 mm in diameter; tunics brown, firm-textured, hidden by the cataphylls. Cataphylls dark-brown, initially unbroken, but often persisting and then becoming fragmented. Foliage leaf single, often emergent at flowering time and shorter than the stem at anthesis, becoming much taller, 9–15 mm wide, linear, channelled, becoming flat above; sheathing leaves 2(4), more or less dry, imbricate, reddish-brown. Stem simple. Rhipidium single and terminal; spathes green, often flushed red, becoming dry and brown, attenuate, the inner 7–10(12) cm long, the outer 2–3 cm shorter than the inner. Flowers pale- to deep-blue to violet with white to yellow nectar guides on the outer tepals; outer tepals 40–65 mm long, the limb about as long as the claw, spreading 30°–40° below the horizontal, inner 35–45 mm long, more or less erect. Filaments 9–15 mm long, united in the lower half, anthers 8–12 mm long. Ovary 15–20 mm long, cylindric and trigonous; style branches 15–20 mm long, crests 10–20 mm long. Capsule 25–35 mm long, nearly cylindric.

Zambia. N: Mbala Distr., Lumi R. flats, Nkali Dambo, 1740 m, 22.ix.1966, *Richards* 31454 (K). W: Mwinilunga Distr., Kakema R., 25.viii.1930, *Milne-Redhead* 965 (PRE). C: Serenje, 27.ix.1961, *Fanshawe* 6727 (NDO; SRGH). E: Nyika Plateau, 2.i.1959, *Robinson* 3012 (K; M; PRE; SRGH). **Zimbabwe**. C: Harare, ix.1919, *Eyles* 1791 (K; PRE; SAM; SRGH). E: Chimanimani (Melsetter), Markham's Kloof, 27.ix.1950, *Crook* M151 (K; LISC; MO; PRE; SRGH). **Malawi**. N: Nyika Plateau, 24.x.1958, *Robson & Angus* 326 (BM; BR; K; LISC; SRGH). C: Dedza, c. 1500 m, 13.ix.1946, *Brass* 17632 (K; MO; SRGH). S: Chambe Plateau, Mt. Mulanje, 3.xi.1956, *Newman & Whitmore* 653 (BR; K; SRGH; WAG). **Mozambique**. N: Amaramba near Mandimba, 8.x.1942, *Mendonça* 666 (LISC). T: between Furancungo and Angónia, 22.ix.1941, *Torre* 3362 (LISC). Z: Gurué Mt., E of the Namuli Peaks, c. 1200 m, 6.xi.1967, *Torre & Correia* 15923 (LISC; MO).

Also in Angola, Zaire, Cameroon, Nigeria, Tanzania, Uganda and Ethiopia, this is one of the most widespread species in the genus. Favouring permanently moist sites; generally flowering at the end of the dry season or early in the wet season, mostly in October to December in central Africa.

This species is distinctive not only in its early flowering but in the reddish-brown, dry cataphylls and sheathing leaves, and the foliage leaf usually only beginning to emerge at anthesis; plants flowering after mid November usually have a more fully developed foliage leaf.

5. GYNANDRIRIS Parl.

Gynandriris Parl., Nuov. Gen. & Sp.: 49 (1854). —Goldblatt in Bot. Not. **133**: 247 (1980).

    Iris section *Gynandriris* (Parl.) Benth. & Hook., Gen. Pl. **3**: 687 (1883).

    Iris subgenus *Gynandriris* (Parl.) Lawrence in Gentes Herb. **8**: 366 (1953).

Perennial herbs with tunicate corms, aerial parts dying back annually. Leaves several to many, the lower 2–3 entirely sheathing and membranous (cataphylls); foliage leaves 1 or

2, linear, bifacial and channelled; sheathing leaves on the upper stem few. Stem simple or branched, the nodes bearing sessile lateral inflorescences. Flowers in umbellate clusters (rhipidia); rhipidia 1–several per branch, sessile except the terminal one, enclosed in a pair of opposed leafy bracts (spathes) concealing the buds; spathes lightly ribbed with alternating bands of green and transparent tissue. Flowers more or less sessile, *Iris*-like, radially symmetrical, usually blue to violet or purple or white, often with yellow nectar guides on the outer tepals. Tepals free, the outer larger and clawed, the limbs spreading, the inner erect or spreading. Filaments united below around the style; anthers oblong to linear, appressed to the style branches. Ovary elongate-cylindric, nearly sessile, included in the spathes, fertile only in the lower part, forming a sterile tube above; style filiform below, dividing into 3 flat, more or less diverging petaloid branches, forked apically into 2 crests, stigma transverse and abaxial below the crests (or terminal). Capsules cylindric, the walls more or less translucent.

A genus of 9 species, 1 in south tropical Africa, 7 in southern Africa and 2 in the Mediterranean and Middle East.

Gynandriris simulans (Baker) R.C. Foster in Contrib. Gray Herb., No. 114: 40 (1936). —Goldblatt in Bot. Not. **133**: 252 (1980). Tab. 7 fig. A. Type from South Africa (Transvaal).
  Moraea simulans Baker, Handb. Irid.: 58 (1892). Type as above.
  Moraea cladostachya Baker, Handb. Irid.: 58 (1892). Type from South Africa (Cape).
  Moraea burchellii Baker, Handb. Irid.: 57 (1892). Type from South Africa (Cape).
  Helixyra simulans (Baker) N.E. Br. in Trans. Roy. Soc. S. Africa **17**: 350 (1929); in Fl. Pl. S. Africa **16**: pl. 623 (1936).
  Helixyra cladostachya (Baker) N.E. Br. in Trans. Roy. Soc. S. Africa **17**: 349 (1929).
  Helixyra burchellii (Baker) N.E. Br. in Trans. Roy. Soc. S. Africa **17**: 349 (1929).
  Helixyra elata N.E. Br. in Trans. Roy. Soc. S. Africa **17**: 349 (1929). Type from South Africa (Transvaal).
  Helixyra mossii N.E. Br. in Trans. Roy. Soc. S. Africa **17**: 350 (1929). Type from South Africa (Transvaal).
  Helixyra spicata N.E. Br. in Trans. Roy. Soc. S. Africa **17**: 349 (1929). Type from South Africa (Cape).
  Helixyra propinqua N.E. Br. in Trans. Roy. Soc. S. Africa **17**: 349 (1929). Type from South Africa (Transvaal).
  Gynandriris mossii (N.E. Br.) R.C. Foster in Contrib. Gray Herb., No. 114: 40 (1936).
  Gynandriris burchellii (Baker) R.C. Foster in Contrib. Gray Herb., No. 114: 40 (1936).
  Gynandriris cladostachya (Baker) R.C. Foster in Contrib. Gray Herb., No. 114: 40 (1936).
  Gynandriris elata (N.E. Br.) R.C. Foster in Contrib. Gray Herb., No. 114: 40 (1936).
  Gynandriris spicata (N.E. Br.) R.C. Foster in Contrib. Gray Herb., No. 114: 41 (1936).
  Gynandriris propinqua (N.E. Br.) R.C. Foster in Contrib. Gray Herb., No. 114: 40 (1936).

Plants (10)14–40(45) cm high. Corms 1.5–2 cm in diameter; tunics dark-brown, coarsely fibrous; cormlets often present in leaf axils. Foliage leaves 1–2, exceeding the stem, 2–4 mm wide, linear, channelled. Stem either contracted and sometimes with several basal branches, or extended with abbreviated axillary branches and sometimes with longer basal branches. Spathes usually dry at flowering time, occasionally green below, the inner (25)30–55 mm long, the outer slightly shorter. Flowers pale blue-lilac with darker speckles on the tepal limbs, and with yellow nectar guides at the base of the outer tepal limbs; outer tepals (17)20–28 mm long, limb nearly equal to or slightly longer than the claw, and up to 8 mm wide, spreading to reflexed; inner tepals 15–19 × 3–4 mm, spreading to reflexed. Filaments 5–6 mm long, united basally for 1–1.5 mm; anthers (4)5–8 mm long. Style branches 8–12 mm long, crests 3–6 mm long. Capsule 13–20 mm long, oblong. Seeds large, blackish, angular.

Botswana. SE: c. 16 km N of Kanye, 27.ix.1964, *Leach, Bayliss & Lamont* 12478 (K; SRGH).
Also in South Africa (Transvaal, Orange Free State, Cape Province) and Lesotho. In xeric grassland and bush, usually in stony ground, especially in disturbed sites; flowering in August and September.
The flowers open between 4 and 5 pm. and fade soon after nightfall. This species is distinguished by the tepals being mottled purple on a paler background.
An early record from Zimbabwe [Bulawayo, viii.1876, *Klingberg* s.n. (S)] requires confirmation.

Tab. 7. A. —GYNANDRIRIS SIMULANS, flower shoot (× 1), detail of flower with tepals removed showing long tubular ovary, style branches and stamens, from *Goldblatt* s.n. Drawn by J.E. Klein. B. —HOMERIA PALLIDA, habit (× 1), from *Goldblatt* 4677. Drawn by Margo Branch.

6. HOMERIA Vent.

Homeria Vent., Dec. Pl. Nov.: 2 (1808). —Goldblatt in Ann. Missouri Bot. Gard. **68**: 431 (1981).

Perennial herbs with tunicate corms, aerial parts dying back annually; corm tunics of blackish coarse wiry fibres. Leaves several to many, the lower 2–3 entirely sheathing and membranous (cataphylls); foliage leaves solitary (2–several in species outside the Flora Zambesiaca area), linear, without a midrib, channelled, bifacial. Flowering stems of several internodes, usually branched, with entirely sheathing bract-like leaves at the nodes, the sheaths closed. Inflorescences consisting of single terminal rhipidia; spathes green, coriaceous, the inner longer than the outer. Flowers pedicellate, actinomorphic, more or less stellate, mostly yellow or salmon-pink, tepal markings when present usually on all tepals; tepals free, unguiculate, the limbs spreading. Filaments united in a column around the style, sometimes free near their apices; anthers linear, appressed to the style branches. Ovary cylindric to ovoid and trigonous. Style slender, dividing at the apex of the filament column into 3 narrow, compressed branches; the branches stigmatic at the apices, sometimes bilobed or with paired short subterminal appendages. Capsules coriaceous, often included in the spathes until maturity. Seeds irregularly angled.

A genus of 32 species mostly of the winter rainfall area of South Africa (Cape Province), also in Lesotho, Botswana and Namibia.

Homeria pallida Baker, Handb. Irid.: 75 (1892). —Goldblatt in Ann. Missouri Bot. Gard. **68**: 445 (1981). TAB. **7** fig. B. Type from South Africa (Cape Province).
  Moraea glauca J.M. Wood & M.S. Evans in J. Bot. **35**: 352 (1897). Type from South Africa (Natal).
  Homeria glauca (J.M. Wood & M.S. Evans) N.E. Br. in Trans. Roy. Soc. S. Africa **17**: 350 (1929).
  Homeria townsendii N.E. Br. in Trans. Roy. Soc. S. Africa **17**: 351 (1929). Type from South Africa (Transvaal).
  Homeria humilis N.E. Br. in Trans. Roy. Soc. S. Africa **17**: 351 (1929). Type from South Africa (Transvaal).
  Homeria mossii N.E. Br. in Trans. Roy. Soc. S. Africa **17**: 351 (1929). Type from South Africa (Transvaal).
  Homeria pura N.E. Br. in Trans. Roy. Soc. S. Africa **17**: 351 (1929). Type from South Africa (Orange Free State).

Plants (10)20–40 cm high. Corms 1–2 mm in diameter; tunics of coarse black fibres. Foliage leaf solitary, much exceeding the stem in length and 5–15 mm wide, often bent and trailing, usually sheathing the lower part of the stem, often somewhat glaucous; cauline leaves entirely sheathing and similar to the inflorescence spathes. Stem usually branched. Rhipidia solitary; spathes 3.5–5(6) cm long, the inner slightly longer than the outer. Flowers yellow (rarely pink), darker toward the centre and with minute dark-greenish spots near the bases of the tepals, lasting for one day; tepals 18–24(27) mm long, claws c. 3 mm long and appressed to the filament column, limb 5–8(12) mm wide. Filaments 5–7 mm long, partly united into a slender column, free and diverging in the upper 1.5–2 mm; anthers straight and diverging, becoming arched after anthesis, 4–6 mm long, exceeding the style branches. Ovary c. 10 mm long, style branches 5–6 mm long and c. 1 mm wide with crests c. 1 mm long, stigma bilobed. Capsules 1–2 cm long, oblong. Seeds angular.

Botswana. SE: near Gaborone Dam, ix.1967, *Lambrecht* 324 (K; SRGH). Mochudi, *Rogers* 6229 (Z).
Also in central Namibia, South Africa (N Cape Province, Orange Free State, Transvaal, Natal) and Lesotho. In open grassland, sometimes in vleis, or in rocky sites; flowering in August to October, usually before the rains.
Known to be toxic to stock, especially in the early stages of growth; fatal if eaten in any quantity.

7. SAVANNOSIPHON Goldblatt & Marais

Savannosiphon Goldblatt & Marais in Ann. Missouri Bot. Gard. **66**: 849 (1979).

Perennial herbs with small globose corms, aerial parts dying back annually; corm tunics membranous to fibrous. Leaves few, the lower 2–3 entirely sheathing (cataphylls); the foliage leaves lanceolate, slightly plicate, the lowermost basal and attached to the stem near ground level, the upper foliage leaves cauline and smaller. Stem compressed and winged, unbranched. Inflorescence a spike; bracts green, fairly large. Flowers white, actinomorphic, with a long perianth tube; tepals subequal, or the uppermost tepal somewhat hooded. Stamens included in the perianth tube, unilateral and pressed against the adaxial part of the tube. Style included in the perianth tube, filiform, 3-branched, the branches deeply divided and recurved. Capsules oblong, coriaceous, showing the outline of the seeds. Seeds several per locule, globose, hard, more or less smooth.

A genus of one species of south tropical Africa.

Although it has for some time been included in *Lapeirousia* because of its deeply divided style branches, the genus seems to be taxonomically isolated. Its round-based corms with membranous to fibrous tunics differ from the flat-based campanulate corms of *Lapeirousia* with hard, often woody tunics. Nevertheless, it seems to be most closely related to this widespread African genus.

Savannosiphon euryphyllus (Harms) Goldblatt & Marais in Ann. Missouri Bot. Gard. **66**: 849 (1979).
    TAB. **8** fig. A. Type from Tanzania.
        Lapeirousia euryphylla Harms in Bot. Jahrb. Syst. **28**: 366 (1900). —Geerinck et al. in Bull. Soc. Roy. Bot. Belgique **105**: 346 (1972). Type as above.
        Acidanthera euryphylla (Harms) Diels in Engler & Prantl, Nat. Pflanzenf. ed. 2, **15a**: 492 (1930).

Plants 15–30 cm high. Corms c. 1 cm in diameter; tunics of medium to coarse fibres. Leaves 3–5, the lower 1–2(3) broadly laminate, about as long as or slightly longer than the stem, lanceolate, plicate; the upper leaves shorter and partly to almost entirely sheathing. Stem erect, compressed and 2-winged, unbranched. Spike 1–4(6)-flowered; bracts 4–6(20) cm long, green, the inner shorter than the outer. Flowers white, night-blooming, odourless, actinomorphic; perianth tube 8–12.5 cm long, nearly cylindrical; tepals 4–5 cm long, lanceolate, acute, the uppermost tepal sometimes larger and somewhat hooded. Filaments inserted 15 mm below the mouth of the perianth tube; anthers c. 10 mm long, with prominent acute appendages, entirely included or partly emerging from the perianth tube. Style dividing near the apex of the anthers, usually the apices of the style branches exserted from the perianth tube. Capsules c. 15 mm long.

Zambia. N: Lake Young, Ishiba (Shiwa) Ngandu, c. 1350 m, 17.i.1959, *Richards* 10717 (K). W: Solwezi Distr., 10 km W of Mutanda, 19.i.1975, *Brummitt, Polhill & Chisumpa* 13847 (K; NDO; SRGH; WAG). C: c. 9 km E of Lusaka, 16.i.1956, *King* 272 (K). E: Cikhasu (Kundu), Luangwa Valley, near river, 13.i.1966, *Astle* 4374 (NDO; SRGH). **Malawi**: N: Mzuzu, Marymount, 20.ii.1976, *Pawek* 10861 (MO; SRGH; WAG). C: near Chinthembwe (Chintembe) Mission, road into Ntchisi Forest Reserve, 14.i.1967, *Hilliard & Burtt* 4497 (MAL; SRGH). S: Ntcheu Distr., Lake View Mission, 17.i.1972, *Salubeni* 1719 (MAL). **Mozambique**. N: Massangulo, i.1933, *Gomes e Sousa* 1252 (COI; K; LMA; WAG).
    Also in Tanzania and Zaire (Shaba). In woodland, often associated with termite mounds. Flowering in January and February.

8. LAPEIROUSIA Pourr.

Lapeirousia Pourr. in Mém. Acad. Sci. Toulouse **3**: 79–82 (1788). —Baker in F.T.A. **7**: 350–355 (1898), including *Anomatheca*. —Goldblatt in Ann. Missouri Bot. Gard. **77**: 430–484 (1990).

Perennial herbs with bell-shaped flat-based corms, aerial parts die back annually; corm tunics of densely compacted fibres or woody. Leaves several, the lower 2–3 membranous and sheathing the stem base (cataphylls); the foliage leaves few or sometimes solitary, plane, or shallowly plicate-corrugate, or terete, the lowermost longest and inserted on the stem near ground level, the upper leaves cauline and progressively smaller. Stem somewhat compressed and angular, sometimes entirely subterranean. Inflorescence panicle-like, or a simple to branched spike, or flowers clustered at ground level; bracts green to membranous, the outer sometimes ridged, keeled, crisped or toothed. Flowers blue, purple, red, white or pink, actinomorphic or zygomorphic; tepals connate into a short to long perianth tube; tepal lobes subequal or unequal. Stamens symmetrically disposed around the style or unilateral and arcuate. Style filiform; style branches usually forked for up to half of their length, sometimes entire or barely bifid. Capsules membranous to coriaceous, more or less globose. Seeds more or less globose.

Tab. 8. A. —SAVANNOSIPHON EURYPHYLLUS, flowering shoot (×$\frac{1}{2}$), dissected flower (× c. 1), from *Pawek* 7972. Drawn by Y. Wilson. B. —LAPEIROUSIA COERULEA, whole plant (×$\frac{1}{2}$), flower (×1), separated stamens and style (× 3), from *Goldblatt & Manning* 8811A. Drawn by J.C. Manning.

A genus of c. 40 species widespread across sub-Saharan Africa, from the SW Cape Province to Nigeria and Ethiopia, with centres of distribution in South Africa (Cape Province) and northern Namibia; 12 species occurring in the Flora Zambesiaca area.

1. Axis contracted above ground level, the inflorescence congested and plants tufted in appearance; floral bracts green and hardly different from the leaves   12. *odoratissima*
   - Axis, including the inflorescence, extending above the ground and relatively lax; floral bracts green to more or less membranous and dry but unlike the foliage leaves   -   -   -   2
2. Perianth tube (20)25–150 mm long (if less than 25 mm then the tepals less than 2 mm wide)   -   -   -   -   -   -   -   -   -   -   -   -   -   -   -   3
   - Perianth tube 1–20(25) mm long (if 20–25 mm then the tepals at least 2 mm wide)   7
3. Perianth tube 100–150 mm long; tepals 6–7 mm wide   -   -   -   -   10. *schimperi*
   - Perianth tube 20–70 mm long; tepals 1.5–5 mm wide   -   -   -   -   -   -   4
4. Tepals at least 15(30) mm long; flower colours in shades of white to cream, rarely purple, with or without markings on the lower tepals   -   -   -   -   -   -   -   11. *littoralis*
   - Tepals 8–14 mm long; flower colours in shades of blue to violet, sometimes white to cream, pink, red, or greenish, but nearly always with contrasting markings on the lower 3 tepals   5
5. Inflorescence a panicle or spike with at least some main branches 3–8-flowered
                                                                            7. *masukuensis*
   - Inflorescence a panicle with the main branches 1(2)-flowered   -   -   -   -   6
6. Perianth white to pale-pink; corm tunics straw-coloured and more or less fibrous and reticulate; perianth tube 25–34(40) mm long, rarely less   -   -   -   -   -   - 9. *bainesii*
   - Perianth blue to violet; corm tunics dark-brown, composed of irregularly broken strips; perianth tube rarely exceeding 22 mm in length   -   -   -   -   -   8. *sandersonii*
7. Perianth tube less than 2 mm long; flowers actinomorphic with stamens symmetrically disposed   -   -   -   -   -   -   -   -   -   -   -   -   - 1. *coerulea*
   - Perianth tube 3–20(25) mm long; flowers zygomorphic with arcuate unilateral stamens   8
8. Perianth tube 15–20(25) mm long   -   -   -   -   -   -   -   -   -   9
   - Perianth tube (3)8–12(15) mm long   -   -   -   -   -   -   -   -   - 10
9. Main inflorescence branches 1(2)-flowered; leaves 1–2 mm wide; branching more or less divaricate   -   -   -   -   -   -   -   -   -   -   -   - 8. *sandersonii*
   - Main inflorescence branches 5–8-flowered; leaves 3–6 mm wide; branches unequal with the main axis straight   -   -   -   -   -   -   -   -   - 7. *masukuensis*
10. Inflorescence a lax panicle with the main terminal branches mostly 3–5-flowered; the dorsal tepal sub-erect to curving forwards over the stamens   -   -   -   -   2. *rivularis*
   - Inflorescence a lax to dense panicle with the main terminal branches 1–5-flowered; the dorsal tepal erect to patent, spreading outwards, thus not hooded over the stamens   -   -   11
11. Leaves ± terete, narrow, elliptic to circular in section, 1–1.5 mm in diameter, without an evident midrib; perianth tube 3–5 mm long and tepals 5–14 mm long; main inflorescence branches 1–2-flowered   -   -   -   -   -   -   -   -   -   -   -   -   -   -   - 12
   - Leaves flat, narrow to broad, 0.5–8(11) mm wide but always with a defined midrib; perianth tube 7–14 mm long and tepals 7–11 mm long; at least some inflorescence branches with more than 2(8) flowers   -   -   -   -   -   -   -   -   -   -   -   -   -   - 13
12. Tepals 5–6 mm long; perianth tube exceeding the bracts by 2–3 mm and only slightly shorter than the tepals   -   -   -   -   -   -   -   -   -   -   - 5. *teretifolia*
   - Tepals 12–14 mm long; perianth tube barely exceeding the bracts and less than half as long as the tepals   -   -   -   -   -   -   -   -   -   -   - 6. *zambeziaca*
13. Plants rarely more than 12 cm high; bracts green at anthesis; branches usually contorted
                                                                            4. *setifolia*
   - Plants (15)20–45 cm high; bracts generally membranous and dry above at anthesis; branches not contorted   -   -   -   -   -   -   -   -   -   -   - 3. *erythrantha*

1. **Lapeirousia coerulea** Schinz in Verh. Bot. Vereins Prov. Brandenburg **31**: 212 (1890). —Baker in F.T.A. **7**: 351 (1898). —Sölch in Merxm., Prodr. Fl. SW. Afrika, fam. 155: 8 (1969). —Goldblatt in Ann. Missouri Bot. Gard. **77**: 445 (1990). TAB. **8** fig. B. Type from Namibia.
   Ixia dinteri Schinz in Mém. Herb. Boissier, No. 20: 14 (1900). Type from Namibia.

Plants (12)15–30 cm high. Corms 12–16 mm in diameter at the base; tunics of light-brown compacted fibres, becoming coarsely to finely fibrous and reticulate. Foliage leaves 2–4; the lowermost inserted at the base of the stem and longer than the rest, reaching to at least the middle of the inflorescence, sometimes shortly exceeding it, 2–4(5) mm wide, more or less linear; the upper leaves decreasing in size above. Stem compressed, 2–3-angled, often many-branched. Inflorescence a lax panicle with ascending branches, the main branches each bearing (2)3–5 sessile flowers; outer bracts 3–4 mm long, green below, membranous apically and becoming completely dry in fruit, inner bracts about as long as the outer. Flowers actinomorphic, more or less stellate, blue to light-purple with a white hastate marking outlined in dark-blue to violet or red in the lower half of each tepal; perianth tube 1–1.5 mm long, cylindric below, widening above, the cylindric part less than 1 mm long; tepals subequal, 7–9.5 × 3–5 mm, lanceolate, spreading, curving upwards toward the apex. Stamens symmetrical around the style; filaments c. 4 mm long, united basally for c. 0.5 mm, erect; anthers 3–3.5 mm long, diverging, incurved after anthesis. Style erect, dividing near the anther apices, style branches c. 1.2 mm long, lightly notched apically. Capsules c. 4 mm long.

Botswana. N: Dobe region, north of Aha Hills near Namibia border, 25.iv.1980, *P.A. Smith* 3496 (MO; PRE; SRGH). SW: Ghanzi Distr., Kuke, pan on Farm 102, 21.ii.1970, *Brown & Brown* 8723 (C; PRE; SRGH).

Also in Namibia. Typically in seasonally wet, low-lying or poorly drained sites, in pans and grassland, or rock outcrops; flowering in December to March.

2. **Lapeirousia rivularis** Wanntorp in Svensk Bot. Tidskr. **65**: 53–56 (1971). —Goldblatt in Ann. Missouri Bot. Gard. **77**: 448 (1990). TAB. **9** fig. A. Type from Namibia.

Plants 30–45 cm high. Corms 12–22 mm in diameter at the base; tunics dark-brown to blackish, coarsely fibrous. Foliage leaves 3–5, the lower 2 more or less basal, channelled for at least half their length, reaching to about the base of the inflorescence, 2–3 mm wide, linear and unifacial above, firm-textured, midrib lightly raised; the upper leaves subtending the inflorescence branches, becoming shorter above. Stem weakly compressed and 3–4-angled, usually laxly branched. Inflorescence a rounded to elongate panicle, the ultimate branches bearing (1)3–5 flowers, laxly arranged; the outer bracts 4–6(9) mm long, green below, membranous and reddish above, apically dry and brown-tipped in bud, later becoming dry for two-thirds of their length, the inner bracts about as long as the outer and entire or apically forked. Flowers zygomorphic, pale blue-mauve, the lower 3 tepals each with a white to cream-coloured nectar guide outlined in deep-violet, sometimes with a median red streak, the reverse of the tube bluish; perianth tube 7–9 mm long, c. 1.3 mm wide at the base and 3 mm wide at the mouth, narrowly campanulate, weakly curved in the upper third; tepals subequal, (7)10–13 × 4–5 mm, lanceolate, the margins somewhat undulate, the dorsal tepal suberect to arching over the stamens, the others directed forwards, the lower 3 nearly horizontal, joined for c. 1 mm more than the dorsal, sometimes each with a tooth-like callus in the median lower half. Stamens unilateral and arcuate; filaments 8.5 mm long, exserted c. 5 mm from the tube; anthers 3.5–4 mm long. Style dividing near the anther apices; style branches c. 3 mm long, each shortly to deeply divided (rarely entire). Capsules 4–5 mm long.

Zambia. B: Machili, 24.xii.1960, *Fanshawe* 6015 (BR; K; LISC; SRGH). W: Mwinilunga Distr., Kalene Hill, 15.xii.1963, *Robinson* 6051 (M; WAG). C: Lusaka, Church Road, in grounds of Evelyn Hone College, 11.i.1986, *Goldblatt* 7537 (E; K; MO; NBG; PRE; S; WAG; WIND). S: Mazabuka, Ridgeway Road, 15.xii.1931, *Trapnell* 588 (BR; K; PRE; Z).

Also in Namibia and Angola. Mostly in moist or wet localities, (dambos or pan margins), but also in rocky grassland.

Easily confused with *L. erythrantha* in which the dorsal tepal is reflexed in the fully open flower so that it lies in the same plane as the other tepals, and is thus not inclined over the stamens as in *L. rivularis*.

3. **Lapeirousia erythrantha** (Klotzsch ex Klatt) Baker in J. Linn. Soc., Bot. **16**: 155 (1878); in F.T.A. **7**: 351 (1898). —Rendle in J. Linn. Soc., Bot. **40**: 210 (1911). —Geerinck et al. in Bull. Soc. Roy. Bot. Belgique **105**: 335 (1972), excluding var. *teretifolia* and var. *welwitschii*. —Tredgold & Biegel, Rhod. Wild Fl.: 12, pl. 7 (1979). —Goldblatt in Ann. Missouri Bot. Gard. **77**: 451 (1990). TAB. **9** fig. B. Type: Mozambique, Cabaceira, *Peters* s.n. (B, holotype).
Ovieda erythrantha Klotzsch ex Klatt in Peters, Naturw. Reise Mossambique: 516, t.58 (1864). Type as above.
Lapeirousia sandersonii sensu Baker in F.T.A. **7**: 352 (1898) non Baker (1892); sensu Eyles in Trans. Roy. Soc. S. Africa **5**: 332 (1916); sensu Suessenguth & Merxmüller, Contrib. Fl. Marandellas Distr.: 76 (1951).
Geissorhiza briartii De Wild. & T. Durand in Bull. Soc. Roy. Bot. Belgique **39**, 4: 105 (1900). Type from Zaire.
Lapeirousia rhodesiana N.E. Br. in Bull. Misc. Inform., Kew. **1906**: 169 (1906). —Eyles in Trans. Roy. Soc. S. Africa **5**: 332 (1916). —Sealy in Bot. Mag. **172**: t. 349 (1959). —Hepper in F.W.T.A., ed. 2, **3**: 141 (1968). Type: Zimbabwe, Headlands, without date, *Evelyn Cecil* 154 (K, holotype).
Lapeirousia graminea Vaupel in Bot. Jahrb. Syst. **48**: 533 (1912). Type: Mozambique, Dondo (25 Mile Station), 10.iv.1898, *Schlechter* 12238 (B, holotype).
Lapeirousia spicigera Vaupel in Bot. Jahrb. Syst. **48**: 547, 548 (1912). Type from Angola.
Lapeirousia plagiostoma Vaupel in Bot. Jahrb. Syst. **48**: 547 (1912). Type: Mozambique, Station Howesa, rocky slopes, 22.i.1906, *Tiesler* 46 (B, holotype).

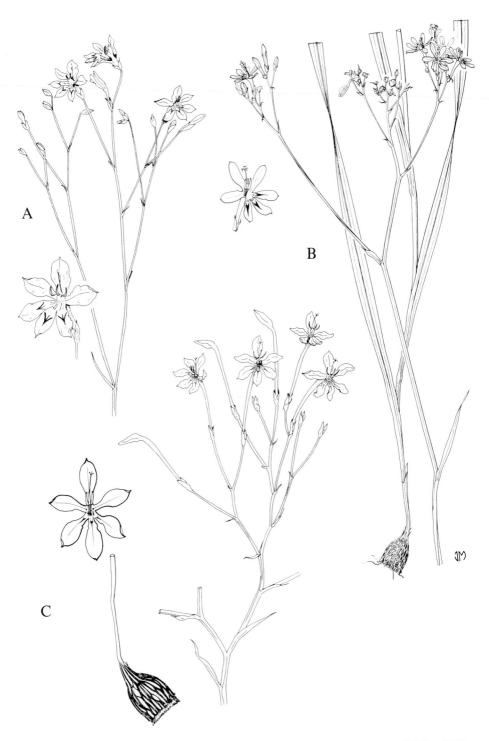

Tab. 9. A. —LAPEIROUSIA RIVULARIS, inflorescence (×½), flower (× 1), from *Goldblatt* 7537.
Drawn by Margo Branch. B. —LAPEIROUSIA ERYTHRANTHA, whole plant (×½), flower
(× 1), from *Goldblatt* 7525. C. —LAPEIROUSIA BAINESII, inflorescence and corm (×½), front
view of flower (× 1), from *Goldblatt & Manning* 8808. Drawn by J.C. Manning.

Lapeirousia erythrantha var. *briartii* (De Wild. & T. Durand) Geerinck et al. in Bull. Soc. Roy. Bot. Belgique **105**: 337 (1972).

Lapeirousia erythrantha var. *rhodesiana* (N.E. Br.) Marais ex Geerinck et al. in Bull. Soc. Roy. Bot. Belgique **105**: 336 (1972).

Plants (15)20–45 cm high. Corms 8–16 mm in diameter at the base; tunics blackish, of densely compacted fibres, the outer layers coarsely fibrous, sometimes becoming finely reticulate and then forming a matted layer. Foliage leaves 3–4, the basal 2–3 leaves one- to two-thirds as long as the stem, the upper ones decreasing in size above, all (2)4–8(11) mm wide, linear to lanceolate, sometimes falcate, only the midrib prominent in living material. Stem compressed and 2-angled below, 3-angled above, branched repeatedly. Inflorescence a several- to many-branched panicle, often more or less umbellate, the ultimate branches with (2)3–6(8) flowers, these often crowded terminally; the outer bracts 3–6 mm long, obtuse, green below, dry and membranous above in bud, becoming entirely dry and brownish in the upper third, the inner bracts usually slightly shorter, acute or bifurcate. Flowers zygomorphic, either blue-violet and the lower tepals each with a hastate white mark outlined in dark-blue to purple, or tepals crimson and then usually unmarked but occasionally the lower tepals each with a white streak in the midline, tepals rarely uniformly white or white with dark markings; perianth tube (6)7–11(14) mm long, slender below, expanded and slightly curved near the throat; tepals subequal, (6)7–11 × 2–3(4) mm, lanceolate to spathulate, acute to obtuse, lying nearly in one plane, the dorsal tepal held apart and reclined, the lower 3 narrow at the base and joined for c. 1 mm. Stamens unilateral; filaments exserted 3–4.5 mm; anthers (2.5)3–4 mm long. Style dividing at mid- to upper-anther level, long; style branches 1–1.5 mm long, forked for a third to half of their length. Capsules 3–4 mm long.

Botswana. N: between Francistown and Nata R., on Maun road, 21.i.1959, *West* 3898 (BR; K; LISC; PRE). **Zambia**. N: Mbala, Uningi Pans, 22.ii.1959, *Richards* 10957 (BR; K; MO). W: Kitwe, Parklands, at the end of Lincoln Ave., 2.iii.1961, *Linley* 85 (K; LISC; MO; SRGH). C: Luanshya, 22.i.1954, *Fanshawe* 711 (BR; K; NDO). E: Chipata (Fort Jameson), Sumbi Hills, 950 m, 3.i.1959, *Robson* 1025 (BR; K; LISC; PRE; SRGH). S: Mazabuka Distr., 6 km from Chirundu Bridge, 6.ii.1958, *Drummond* 5493 (BR; LISC; PRE; SRGH). **Zimbabwe**. N: Darwin Distr., Kandeya Native Reserve, 1000 m, 17.i.1960, *Phipps* 2293 (BR; MO; PRE; SRGH). W: Nyamandhlovu Pasture Station, i.1953, *Plowes* 1540 (K; MO; LISC; SRGH). C: Domboshawa, 16.ii.1958, *Leach* in GHS 83643 (BR; GRA; K; P; SRGH). E: Mutare Distr., Zimunya's Reserve, 25.i.1959, *Chase* 7045 (BR; K; LISC; PRE; SRGH). S: 8 km from Masvingo (Fort Victoria) on the Save (Sabi) road, i.1969, *Goldsmith* 10/69 (K; LISC; MO; PRE; SRGH). **Malawi**. N: Karonga Distr., Sungilo Point, 705 m, 24.ii.1978, *Pawek* 13822 (BR; MO; WAG). C: Dedza Distr., Kangoli Hill, 13.ii.1967, *Salubeni* 563 (K; LISC; PRE; SRGH). S: near Chileka Post Office, 7.i.1986, *Goldblatt* 7524 (MAL; MO; NBG; PRE; WAG). **Mozambique**. N: Nampula, 7.i.1937, *Torre* 1278 (COI). T: Angónia, 2.xii.1980, *Macúacua* 1363 (MO; PRE; WAG). Z: Mocuba Distr., Namagoa Estate, 60 m, xii–i.1943, *Faulkner* 156 (BR; COI; K; P; PRE; S; SRGH). MS: Dombe Distr., above E bank of the Makurupini R., 5 km above the confluence of the Haroni R., 6.i.1969, *Bisset* 13 (K; LISC; LMU; PRE; SRGH).

Also in Angola, Zaire (Shaba), and S and W Tanzania. Usually in shallow soils over rock, on or at the base of granite outcrops, or in lateritic pans or shallow pans in areas of impeded drainage in mopane or miombo woodlands, less often in open grassland. Flowering in mid December to March.

There are two colour forms, a deep maroon-red, occurring in the east of the range and a blue–violet with white and purple nectar guides, which is more widespread. Corms are edible raw or cooked but not a significant source of food today.

4. **Lapeirousia setifolia** Harms in Bot. Jahrb. Syst. **30**: 278 (1902). —sensu Eyles in Trans. Roy. Soc. S. Africa **5**: 332 (1916) non Baker (= *L. erythrantha*). —Goldblatt in Ann. Missouri Bot. Gard. **77**: 456 (1990). Type from Tanzania.

Lapeirousia welwitschii sensu Eyles in Trans. Roy. Soc. S. Africa **5**: 332 (1916) non Baker (= *L. erythrantha*).

Lapeirousia erythrantha var. *setifolia* (Harms) Geerinck et al. in Bull. Soc. Roy. Bot. Belgique **105**: 344 (1972).

Plants small, 5–10(15) cm high. Corms 10–12 mm in diameter at the base; tunics with the inner layers firm and unbroken, becoming decayed and more or less fibrous with age, dark-brown. Foliage leaves 3–7, the lower 2–4 clustered at the stem base and ascending, or ± spreading, the lowermost longest and exceeding the inflorescence, the upper leaves decreasing in size, 0.5–1 mm wide, linear above, usually channelled to above the middle and broader towards the base. Stems irregularly flexuous and somewhat contorted, several- to many-branched, 3–4-angled. Inflorescence a congested pseudopanicle, the ultimate branches 2–4-flowered, the internodes half as long as the bracts; the outer bracts 4–5 mm long, green, often purple-flushed, the inner bracts as long as the outer or slightly shorter, apices dry and apiculate, recurved. Flowers zygomorphic, blue to violet, the lower tepals each with a white and dark-blue mark in the lower midline; perianth tube 8–10 mm long, slender, expanded and curved in the upper 1.5 mm; tepals 7–9 × c. 2 mm, narrowly lanceolate, spreading at right angles to the tube. Stamens unilateral; filaments exserted 2–2.5 mm from the tube; anthers c. 1.5–2 mm long. Style dividing near mid-anther level; style branches c. 1 mm long, divided for about half their length. Capsules c. 2 mm long.

Zimbabwe. E: Nyanga, Matemma, 11.i.1967, *Plowes* 2841 (BR; P; SRGH). W: Matobo, farm Quaringa, c. 1450 m, xii.1953, *Miller* 2004 (B; BR; LISC; PRE; S; SRGH). **Malawi.** N: Nyika Plateau, Chelinda Bridge, c. 2300 m, 29.iii.1970, *Pawek* 3421 (B; K; MAL; MO).
Also in SW Tanzania. In shallow, seasonally waterlogged soil on rock outcrops. Flowering from December to March.
Dwarf plants from the Chimanimani Highlands of eastern Zimbabwe may belong here, but have larger flowers and are therefore more likely to be depauperate specimens of *L. erythrantha*.

5. **Lapeirousia teretifolia** (Geerinck et al.) Goldblatt in Ann. Missouri Bot. Gard. **77**: 457 (1990). Type from Zaire (Shaba).
    Lapeirousia erythrantha var. *teretifolia* Geerinck et al. in Bull. Soc. Roy. Bot. Belgique **105**: 342 (1972). Type as above.

Plants 20–40 cm high. Corms 9–13 mm in diameter; tunics dark-brown to blackish, woody, the outer layers breaking into parallel vertical segments. Foliage leaves 3–4, the lowermost inserted near ground level and longer than the rest, up to half as long, to as long as the stem, with the upper ones decreasing in size above, all rigid, 1–1.5 mm in diameter and more or less terete to elliptic in section, without a discrete midrib. Stems more or less terete below, lightly 3–4 angled above. Inflorescence a rounded to columnar panicle, the main axis usually dominant, the ultimate branches 1–2(3)-flowered; the outer bracts (2.5)3–4 mm long, more or less membranous, transparent below, rust-brown above, or entirely brownish, the inner bracts slightly longer than the outer. Flowers weakly zygomorphic, whitish to pale-lilac, the lower tepals each with a darker blue to violet streak in the lower midline; perianth tube 4–5 mm long, slender, slightly expanded above; tepals 5–6 mm long, subequal, spreading more or less at right angles to the tube, the dorsal tepal held apart from the others, the lower 3 held closely together. Stamens unilateral; filaments exserted 2–2.5 mm; anthers 2–3 mm long. Style dividing at mid-anther level; style branches c. 1 mm long, usually forked for half their length, sometimes less. Capsules c. 4 mm long, globose.

Zambia. N: Mwinilunga Distr., Kalenda Ridge west of Matonchi Farm, 22.i.1938, *Milne-Redhead* 4280 (BR; K; LISC; PRE).
Also in southern Zaire (Shaba) and Angola. In seasonally moist or waterlogged ground, often in lateritic pans. Flowering from February to April.
Distinguished from *L. erythrantha* by the paler-coloured bluish to white or pink flowers, these being smaller in size and typically single on each branch of the inflorescence.

6. **Lapeirousia zambeziaca** Goldblatt in S. African J. Bot. **57**: 226 (1991) nom. nov. pro *L. angolensis* Goldblatt. Type from Angola.
    Lapeirousia angolensis Goldblatt in Ann. Missouri Bot. Gard. **77**: 457 (1990), nom. illegit. non. *L. angolensis* (Baker) R.C. Foster. Type as above.

Plants 24–30 cm high. Corms 13–16 mm in diameter; tunics dark-brown to blackish, composed of hard densely compacted fibres, the outer layers breaking into parallel vertical sections. Foliage leaves 2–3, the lowermost inserted near ground level and longer than the rest, about half as long to as long as the inflorescence, the upper ones decreasing in size above, all rigid, 1–1.5 mm in diameter in the middle and more or less terete to elliptic in section, without a discrete midrib. Stems divaricately branched, weakly compressed, 2-angled below, 3-angled above the branches. Inflorescence a rounded, relatively few-flowered panicle, the main axis usually dominant, the terminal branches 1–2-flowered; the outer bracts (4)5–6 mm long, more or less membranous, green below, rust-brown above, becoming entirely brownish, the inner bracts slightly longer than the outer. Flowers zygomorphic, pale-violet, the lower tepals each with a pale-yellow streak edged in purple on the lower midline; perianth tube 3.3 mm long, slender, slightly expanded above; tepals subequal, 13–14 × 1.5–2 mm, more or less differentiated into claws towards the base, limbs lanceolate (their orientation uncertain), the margins undulate, the dorsal tepal apparently held apart from the others, the lower 3 adjacent tepals united at their base for c. 2 mm and held closely together. Stamens unilateral;

filaments erect, exserted c. 5 mm; anthers 3.5 mm long. Style dividing near the anther apices; style branches c. 1.4 mm long, undivided. Capsules unknown.

Zambia. B: Mongu flood plain, 29.i.1966, *Robinson* 6830 (K).
Also in W Angola. In boggy grassland, probably seasonally inundated; flowering in February to April.
Distinctive in its terete leaves and fairly large flowers with short perianth tubes c. 3.3 mm long, and tepals that much exceed the tube in length.

7. **Lapeirousia masukuensis** Vaupel & Schltr. in Bot. Jahrb. Syst. **48**: 545, 546 (1912). —Goldblatt in Ann. Missouri Bot. Gard. **77**: 461 (1990). Type: Mozambique, Inhambane, Insilva Masuku, 23°30'S, 35°20'E, 10.ii.1898, *Schlechter* 12109 (B, holotype; BR; COI; G; K; P; PRE; SAM).

Plants 40–60 cm high. Corms c. 15 mm in diameter at the base; tunics light-brown, coriaceous with a reticulate surface, the outer layers decaying to become coarsely fibrous. Foliage leaves 4–6, the lowermost inserted near ground level and larger than the rest, usually slightly exceeding the inflorescence, the upper ones progressively smaller above, 3–5 mm wide near the midline, narrowly linear-lanceolate, the midribs prominent. Stems compressed and 2-angled to winged below, 3-angled and lightly winged above the branches. Inflorescence a branched spike or pseudopanicle, the main axis more or less straight and dominant, the major branches forming spikes of 5–9 flowers, the flowers often apically crowded on an axis (1)1.5–2 internodes in length, with the bracts overlapping; the outer bracts 5–7 mm long, green at anthesis, becoming dry and scarious later, and then either pale throughout or brownish above and pale below with brown nerves, the inner bracts about as long as or slightly shorter than the outer. Flowers zygomorphic, either blue to violet or greenish-cream, the lower tepals each with a purple to red and white hastate median streak in the distal half; perianth tube (15)20–25 mm long, cylindric, slightly expanded in the upper 4 mm; tepals subequal, 8–10 × 3 mm, lanceolate, spreading at right angles to the perianth tube and lying in more or less the same plane, the dorsal tepal held apart. Stamens unilateral; filaments exserted 2.5–3 mm; anthers 3.5 mm long. Style dividing opposite the upper third of the anthers; style branches c. 2 mm long, forked for about half their length. Capsules c. 5 mm in diameter.

Zimbabwe. S: Mberengwa (Belingwe), near the Mnene road, 27.ii.1931, *Norlindh & Weimarck* 5202 (BR; PRE; S; SRGH); Bikita Distr., Birchenough Bridge-Masvingo (Fort Victoria) road, 16.v.1962, *Noel* 2427 (K; LISC; SRGH). **Malawi**. N: c. 30 km NW of Rumphi, c. 1400 m, 11.iii.1978, *Pawek* 14048 (K; MAL). **Mozambique**. GI: Quissico, 28.ii.1955, *Exell, Mendonça & Wild* 703 (LISC; SRGH). M: Manhiça, vale do Incomati, 26.iii.1979, *de Koning* 7353 (K; LISC; LMA; LMU).
Also in South Africa (Transvaal). Preferring seasonally wet habitats, either in vleis, marsh edges, or waterlogged flats; flowering in February to April.
Distinguished by its long perianth tube, usually 20–25 mm long, and by the 5–9 flowers apically crowded on the main branches of the inflorescence.

8. **Lapeirousia sandersonii** Baker, Handb. Irid.: 169 (1892); in F.C. **6**: 95 (1896); in F.T.A. **7**: 352 (1898) excl. specimens cited. —Bremek. & Oberm. in Ann. Transvaal Mus. **16**: 409 (1935). —Van Druten, Fl. Pl. Africa **31**: 1226 (1956). —Letty, Wild Fl. Transvaal: 77, t. 37 (1962). —Goldblatt in Ann. Missouri Bot. Gard. **77**: 463 (1990). Type from South Africa (Transvaal). *Lapeirousia bainesii* var. *breviflora* Baker in J. Linn. Soc., Bot. **16**: 156 (1878) nom. nud.

Plants 18–35 cm high. Corms 2.5–3 cm in diameter; tunics dark-brown, coriaceous internally, decaying somewhat irregularly into vertical segments, seldom becoming fibrous and never reticulate. Foliage leaves 2–4, the lowermost longest usually exceeding the inflorescence, the upper leaves progressively shorter, all 2–3 mm wide, more or less linear, firm to rigid, the midrib and lateral veins fairly prominent and closely set. Stems compressed, 2-angled to winged below, 3-angled above the branches and often lightly winged on the angles. Inflorescence much branched, sometimes intricately so, the branches more or less divaricate but unequal and a main axis is usually evident; the branches 1–2(3)-flowered; bracts (4.5)5–8(10) mm long, in bud these are green with brown tips, often flushed purple, in flower these become scarious throughout and almost entirely brown. Flowers zygomorphic, blue to violet, the lower 3 tepals each with a deep-red to purple and white hastate mark in the lower midline; the perianth tube 15–18(20) mm long, slender, but slightly wider in the upper 2–3 mm; tepals 10–11 × 3–3.5 mm, subequal, lanceolate; the dorsal tepal held apart from the others and reflexed, the lower adjacent 3 joined at the base for about 1 mm and forming a lip, when fully open all held in more or less the same nearly-horizontal plane. Stamens unilateral; filaments exserted 3 mm; anthers 3–4 mm long. Style dividing near the apex of the anthers; style branches simple or forked apically, c. 1.5 mm long. Capsules 4–5 × c. 5 mm.

Botswana. SE: west of Gaborone Dam, 1976, *Mott* 928 (SRGH; UCBG); Ngwaketse Reserve, 9 km W of Kanye, 12.ii.1971, *van Rensburg* B4226 (PRE).
Also in South Africa (Transvaal, N Cape Province). In rocky grassland, mostly in hilly country; flowering in December to April.
Probably closely related to *L. erythrantha* but easily distinguished by its long perianth tube, c. 20 mm long, its divaricately branched inflorescence with 1–2 flowers per ultimate branch, and by the dark-brown coriaceous, later brittle, corm tunics.

9. **Lapeirousia bainesii** Baker in J. Bot. **14**: 338 (1876); in F.T.A. **7**: 352 (1898). —N.E. Brown in Bull. Misc. Inform., Kew **1909**: 142 (1909). —Bremek. & Oberm. in Ann. Transvaal Mus. **16**: 409 (1935). —Sölch in Merxm., Prodr. Fl. SW. Afrika, fam. 155: 8 (1969) (excl. syn. *L. otaviensis*). —Goldblatt in Ann. Missouri Bot. Gard. **77**: 469 (1990). TAB. **9** fig. C. Type: Botswana, Kobe Pan "inter Koobie et Norton Shaw Valley," *Baines* s.n. (K, lectotype).
Lapeirousia vaupeliana Dinter in Fedde, Repert. Spec. Nov. Regni Veg. **18**: 436 (1922). —Sölch in Merxm., Prodr. Fl. SW. Afrika, fam. 155: 10 (1969). Type from Namibia.

Plants 30–60 cm high. Corms 13–20 cm in diameter; tunics mid- to dark-brown, coriaceous to cartilaginous internally with the veins sharply outlined, decaying to become more or less fibrous, the fibres wiry and coarse. Foliage leaves 5–7 mm wide, usually slightly longer than the inflorescence, narrowly lanceolate to linear, glaucous, midrib prominent with a lateral vein evident on either side. Stems compressed and 2-winged below, 3–4 angled to winged above, the wings sometimes slightly crisped or serrate. Inflorescence a pseudopanicle with divaricate branching, the ultimate branches with 1(2) flowers, the axes trigonous; the outer bracts 7–10(12) mm long, green below drying before anthesis, becoming membranous dry and light-brown especially above, the apices often darker brown, the inner bracts slightly larger than the outer. Flowers sweetly scented and opening in the mid to late afternoon, zygomorphic, white to cream-coloured, sometimes flushed pale-pink, the perianth tube pale-purple, the lower 3 tepals each usually with a red to brown streak in the lower half and a dark mark at the base, the top of the throat red on the lower side; perianth tube 25–34(40) mm long, cylindric, slightly expanded in the upper 5 mm; tepals 9–12(15) × 3–4 mm, widest in the upper third, subequal or the dorsal tepal slightly larger, held apart and upright to reflexed at right angles to the tube, more or less clawed, the margins undulate, the lower adjacent 3 rarely with a tooth-like median callus. Stamens unilateral; filaments exserted 3.5–5 mm; anthers 4.5–6 mm long. Style dividing between the middle and apex of the anthers or sometimes exceeding them; style branches spreading, undivided or notched apically, c. 2 mm long. Capsules 5–6 mm long.

Caprivi Strip. Bagani (Bagoni) Camp, Caprivi side of river, 19.i.1956, *de Winter & Wiss* 4333 (PRE). **Botswana**. N: Dobe, c. 26 km north of Aha Hills, Namibia border, 13.iii.1965, *Wild & Drummond* 7202 (K; PRE; SRGH). SE: c. 15 km W of Machaneng (Macheng) towards Mahalapye, 23°10'S, 27°20'E, 1.iii.1978, *O.J. Hansen* 3357 (C; K; PRE; SRGH; WAG). **Zimbabwe**. S: c. 15 km N of Beitbridge on Masvingo (Fort Victoria) road, 450 m, 21.iii.1967, *Rushworth* 485 (SRGH).
Also in Namibia and South Africa (Transvaal). In hard flat sandy ground, in open or bush country; flowering mostly in January to April.
Easily recognised by its white to pale-pink, long-tubed flowers and its divaricately branched inflorescence. The corms are edible and an important food in Namibia and perhaps in N Botswana.

10. **Lapeirousia schimperi** (Asch. & Klatt) Milne-Redhead in Bull. Misc. Inform., Kew **1934**: 307 (1934). —Goldblatt in Ann. Missouri Bot. Gard. **77**: 472 (1990). TAB. **10** fig. A. Type from Ethiopia.
Tritonia schimperi Asch. & Klatt in Linnaea **34**: 697 (1866). Type as above.
Anomatheca angolensis Baker in J. Bot. **14**: 337 (1876). Type from Angola.
Acidanthera unicolor Hochst. ex Baker in J. Linn. Soc., Bot. **16**: 160 (1878); in F.T.A. **7**: 359 (1898), nom. superfl. pro *Tritonia schimperi* Asch. & Klatt (1866). Type as for *Lapeirousia schimperi*.
Lapeirousia fragrans Welw. ex Baker in Trans. Linn. Soc. London, ser. 2, Bot. **1**: 272, 273 (1878). Type from Angola.
Lapeirousia cyanescens Welw. ex Baker, Trans. Linn. Soc. London, ser. 2, Bot. **1**: 272 (1878). —Eyles in Trans. Roy. Soc. S. Africa **5**: 332 (1916). —Sölch in Merxm., Prodr. Fl. SW. Afrika, fam. 155: 9 (1969). —Cufodontis in Bull. Jard. Bot. Belg. **42**(3) Suppl. [Enum. Pl. Aethiop. Sperm.]: 1592 (1972). Type from Angola.
Lapeirousia edulis Schinz in Bull. Herb. Boissier, No. **4**, Appendix 3: 49 (1896). Type from Namibia.
Lapeirousia monteiroi Baker in F.T.A. **7**: 355 (1898), nom. illegit. superfl. pro *A. angolensis* Baker.

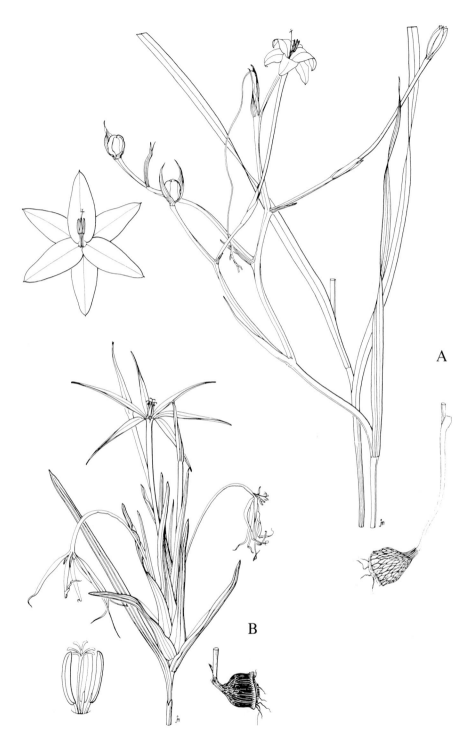

Tab. 10. A. —LAPEIROUSIA SCHIMPERI, whole plant (×$\frac{1}{2}$), front view of flower (× 1), from *Goldblatt & Manning* 8831. B. —LAPEIROUSIA ODORATISSIMA, habit and corm (×$\frac{1}{2}$), detail of stamens and style (× 3), from *Goldblatt* 8824. Drawn by J.C. Manning.

Lapeirousia porphyrosiphon Baker in F.T.A. **7**: 353 (1898). —N.E. Br. in Bull. Misc. Inform., Kew **1909**: 142 (1909). —Eyles in Trans. Roy. Soc. S. Africa **5**: 332 (1916). —Martineau, Rhod. Wild Fl.: 20, pl. 2 (1954). Type: Botswana, Kalahari Desert near Mamunwe, 26.ii.1897, *E.J. Lugard* 238 (K, holotype).
Lapeirousia uliginosa Dinter, Veg. Veldk. Deutsch. SW. Afrikas: 13, 14 (1912). Type unknown (but based on plants from Namibia).
Lapeirousia dinteri Vaupel in Bot. Jahrb. Syst. **48**: 544 (1912). Type from Namibia.
Lapeirousia angolensis (Baker) R.C. Foster in Contrib. Gray Herb., No. 114: 48 (1936).

Plants (20)30–80 cm high. Corms 18–22 mm in diameter; tunics of compacted fibres, light- to dark-brown, the outer layers becoming loosely fibrous and reticulate. Foliage leaves 3 or more, the lower 2 longer than the rest and usually slightly exceeding the inflorescence, linear, the upper leaves decreasing in size and becoming bract-like, 5–10(15) mm wide, narrowly lanceolate, the midrib lightly raised. Stem terete below to nearly square in section and 4-angled to 4-winged above, laxly branched. Inflorescence a lax pseudopanicle, the ultimate branches with 1–3 sessile flowers; bracts subequal, (10)20–35(45) mm long, green becoming membranous above, or completely dry and papery and light- to dark-brown, apices dark-brown. Flowers opening in the evening and then sometimes scented, zygomorphic, white to cream-coloured, rarely pale-violet, sometimes whitish flowers become lilac on drying especially on the tube; perianth tube 10–14(15) cm long, straight, cylindric; tepals subequal, 18–22 × 6–7 mm, lanceolate, extended more or less at right angles to the tube. Stamens unilateral; filaments erect, exserted 5–7 mm from the perianth tube; anthers 6–7 mm long, cream. Style dividing near to or up to 3 mm beyond the anther apices, style branches c. 2 mm long, forked for c. one-third of their length. Capsules 8–12 mm long.

Caprivi Strip: Okavango River, 19 km N of Shakawe on the Botswana border, 16.iii.1965, *Wild & Drummond* 7093 (K; LISC; M; PRE; SRGH). **Botswana**. N: Khwai/Maxwe Road, Moremi Wildlife Reserve, 16.iii.1977, *P.A. Smith* 1936 (BR; K; SRGH). **Zambia**. B: Masese, 9.i.1961, *Fanshawe* 6098 (BR; K; NDO; SRGH). S: Livingstone, 20.i.1929, *Grant* 4507 (MO; PRE). **Zimbabwe**. N: Hurungwe (Urungwe) Distr., Mensa Pan, c. 16 km ESE of Chirundu Bridge, 29.i.1958, *Drummond* 5346 (BM; BR; K; LISC; PRE; SRGH). W: Hwange (Wankie) National Park, Shapi Camp, 27.ii.1967, *Rushworth* 269 (BR; K; LISC; PRE; SRGH). C: near Harare (Salisbury), 20.i.1929, *Young in Moss* 17315 (K). **Malawi**. N: Nyika Plateau, cult. Missouri Bot. Gard., *la Croix in Goldblatt* s.n. (MO).
Also in Namibia, Angola, Tanzania, Kenya, Ethiopia and Sudan. In locally moist sites in arid country, often in washes, dambos, stream sides and seasonal marshes, as well as in damp grassland; flowering in December to March.
The large white flower with a perianth tube 10–15 cm long and the laxly branched inflorescence make this species easily recognisable.
The corms are edible, and eaten raw or roasted in northern Namibia.

11. **Lapeirousia littoralis** Baker in Trans. Linn. Soc. London, ser. 2, Bot. **1**: 273 (1878). —Goldblatt in Ann. Missouri Bot. Gard. **77**: 476 (1990). Type from Angola (Namibe).

Plants (5)10–35 cm high. Corms campanulate, 10–14 mm in diameter at the base; tunics brown, woody, the outer layers breaking irregularly, rarely becoming fibrous, the basal margin crenate to denticulate. Leaves few to several, 1.5–4 mm wide, linear, lightly corrugated, the lowermost inserted on the stem at or just below the ground and longer than the rest, equalling the inflorescence in length, rarely somewhat longer, ascending to falcate or trailing, the upper leaves shorter. Stem simple or branched, more or less terete to weakly angled below the nodes; the branches either crowded below, or laxly arranged. Inflorescence comprising 1–several lax to fairly congested spikes of 5–12 flowers; the outer bracts green, 10–20(25) mm long, weakly keeled, the inner bracts smaller, apically forked, becoming membranous. Flowers with a strong sweet scent, zygomorphic, white to cream-coloured, greenish-yellow, or light purplish-brown; perianth tube more or less dimorphic, slender below, curving outwards and expanded above, (20)30–45(70) mm long, the upper part c. 6 mm long; tepals subequal, 13–30 × 1.3–2.5(3) mm, narrowly lanceolate to linear-filiform, acute to attenuate, spreading equally at right-angles to the tube, vertical. Stamens unilateral; filaments exserted 2–3 mm; anthers 4–5 mm long. Style dividing between the base and middle of the anthers; style branches forked for c. one-third of their length, usually tangled in the anthers. Capsules 6–8(11) mm long.

Easily recognised by the long perianth tube, the narrow tepals and plicate (corrugated) basal leaf. Two subspecies are recognised.

Perianth tube (25)30–45(70) mm long; tepals 18–30 × 1.3–2 mm; bracts 10–18(23) mm long
subsp. *caudata*
Perianth tube 28–35 mm long; tepals 13–15 × 2–3 mm; bracts 15–20(25) mm long
subsp. *littoralis*

Subsp. littoralis
Lapeirousia burchellii Baker, Handb. Irid: 171 (1892); in F.C. **6**: 93, 94 (1896). Type from South Africa (Cape Province).
Lapeirousia ramosissima Dinter in Fedde, Repert. Spec. Nov. Regni Veg. **29**: 255 (1931). —Sölch in Merxm., Prodr. Fl. SW. Afrika, fam. 155: 10 (1969). Type from Namibia.
Lapeirousia streyii Suess. in Mitt. Bot. Staatssamml. München **1**(3): 88 (1951). Type from Namibia.
Lapeirousia caudata subsp. *burchellii* (Baker) Marais & Goldblatt in Contrib. Bolus Herb., No.4: 30 (1972).

Botswana. SE: Ngwaketse Reserve, Morapedi Ranch, 25°38'S, 25°03'E, 15.ix.1978, *O.J. Hansen* 3455 (C; K).
Also in Angola (Namibe), Namibia and South Africa (Cape Province, Transvaal). Usually in sandy, well drained soils, sometimes on dunes; mostly flowering in September and October.

Subsp. caudata (Schinz) Goldblatt in Ann. Missouri Bot. Gard. **77**: 478 (1990). Type from Namibia.
Lapeirousia caudata Schinz in Verh. Bot. Vereins Prov. Brandenburg **31**: 213 (1890). —Baker, Handb. Irid: 172 (1892); in F.T.A. **7**: 352 (1898). —Eyles in Trans. Roy. Soc. S. Africa **5**: 332 (1916). —Sölch in Merxm., Prodr. Fl. SW. Afrika, fam. 155: 8 (1969). Type from Namibia.
Lapeirousia delagoensis Baker, Handb. Irid: 171 (1892); in F.C. **6**: 94 (1896). —Schinz & Junod in Mém. Herb. Boissier, No. **10**: 30 (1900). Type: Mozambique, Maputo (Delagoa Bay, Lourenço Marques), *H. Bolus* 7618 (K, lectotype; BOL; G; SAM).
Lapeirousia lacinulata Vaupel in Bot. Jahrb. Syst. **48**: 546 (1912). Type: Zambia, Katanino (Katinina) Hills, xii.1907, *Kassner* 2170 (B, holotype; BM; BR; E; HBG; K; P; Z).

Plants 20–30 cm high, usually with a few long branches produced from near the base. Inflorescence a spike of 8–12 flowers; bracts 10–18(23) mm long. Flowers with a perianth tube (25)30–45(70) mm long; tepals (18)25–30 mm long and 1.3–2 mm wide, usually nearly filiform especially when dry.

Caprivi Strip. Mpilila Island, 12.i.1959, *Killick & Leistner* 3325 (PRE; SRGH; WIND).
Botswana. N: sandy floodplain of Kwando R., 12.i.1978, *P.A. Smith* 2222 (K; PRE; SRGH). SW: Kgalagadi, 120 km WNW of Hukuntsi on track to Ncojane, 13.iii.1979, *Skarpe* 333 (K; PRE; SRGH; UCBG). **Zambia**. B: c. 9 km N of Senanga, 1.viii.1952, *Codd* 7329 (BM; BR; K; MO; PRE; S; SRGH). N: Lake Mweru, 13.xi.1957, *Fanshawe* 3941 (BR; K; NDO; P). C: Serenje Distr., Kanona, xii.1968, *Williamson* 1337 (K; MO; PRE; SRGH). S: Mazabuka, Ridgeway Road, 2.xii.1931, *Trapnell* 538 (BR; K; PRE). **Zimbabwe**. W: Shangani R., NE of Bulawayo, 7.i.1898, *Rand* 229 (BM). C: Harare (Salisbury), xi.1919, *Eyles* 1885 (K; PRE; SAM; SRGH). **Mozambique**. M: Baia de Maputo (Delagoa Bay), Maputo (Lourenço Marques), 30.xi.1897, *Schlechter* 11540 (B; BR; G; P; PRE; SAM).
Also in Namibia. In sandy soils usually near water, on river sandbanks, lake shores, sandy floodplains and pan margins; mostly flowering in November to March.

12. **Lapeirousia odoratissima** Baker in Trans. Linn. Soc. London, ser. 2, Bot. **1**: 273 (1878); in F.T.A. **7**: 354 (1898). —Eyles in Trans. Roy. Soc. S. Africa **5**: 332 (1916). —Suessenguth & Merxmuller, Contrib. Fl. Marandellas Distr.: 76 (1951). —Martineau, Rhod. Wild Fl.: 20 (1954). —Sölch in Merxm., Prodr. Fl. SW. Afrika, fam. 155: 10 (1969). —Geerinck et al., Bull. Soc. Roy. Bot. Belgique **105**: 344 (1972). —Tredgold & Biegel, Rhod. Wild Fl.: 12, pl. 7 (1979). —Goldblatt in Ann. Missouri Bot. Gard. **77**: 480 (1990). TAB. **10** fig. B. Type from Angola.
Lapeirousia congesta Rendle in J. Linn. Soc., Bot. **30**: 435 (1895). —Baker in F.T.A. **7**: 354 (1897). Type from Tanzania.
Lapeirousia stenoloba Vaupel in Bot. Jahrb. Syst. **48**: 548 (1912). Type from Namibia.
Lapeirousia juttae Dinter, Die Vegetabilische Veldkost Deutsch-Südwest-Afrikas: 13 (1912). Type unknown but based on plants from Namibia.

Plants 10–18(25) cm high, with upper internodes condensed and leaves densely clustered. Corms 15–20 mm in diameter at the base; tunics brown, woody, the outer breaking irregularly, rarely becoming fibrous when decayed. Leaves few, often only 2, usually very similar to the bracts and differing mainly by their position, the lowermost inserted below the ground, and up to twice as long as the bracts, up to 30 cm long and 3–5 mm wide, linear, plicate. Stem basal internode up to 8 cm long, usually arising from just below ground level, the upper internodes contracted, rarely 5–10 mm long; the aerial stem 2–5(10) cm long, simple or with a few short branches. Inflorescence of 1 or more congested spikes, more or less umbellate in appearance, flowers 3–6 per branch; the outer bracts green, 6–15 cm long, lanceolate, lightly plicate, the inner bracts about one-third shorter than the outer and more or less membranous. Flowers usually strongly scented especially in the evening, actinomorphic, hypocrateriform, white to ivory in colour; perianth tube 10–14 cm long, cylindric; tepals 35–40 × 4–5 mm, narrowly lanceolate, widest in the lower third, attenuate, extended horizontally or sometimes slightly drooping. Stamens symmetrically disposed; filaments exserted 1.5–2 mm; anthers 6–8 mm long. Style usually dividing opposite the anther apices, or sometimes shorter; style branches 3–4 mm long, forked for half their length. Capsules 15–18 mm long, concealed in the bracts.

Zambia. B: Barotseland, Moufu, 22.xii.1965, *Robinson* 6749 (WAG). N: Mbala Distr., road to Inono Village, c. 1500 m, 18.i.1955, *Richards* 4142 (BR; K). W: Mwinilunga Distr., S of Samuteba on Solwezi-Mwinilunga road, 19.i.1975, *Brummitt, Chisumpa & Polhill* 13875 (K; NDO; SRGH). C: Luangwa Valley near Mupamadzi River, 600 m, 15.i.1966, *Astle* 4415 (K; SRGH). S: Machili (?Forest Reserve), 22.xii.1960, *Fanshawe* 5994 (K; NDO; SRGH). **Zimbabwe**. W: Nkayi Distr., Gwampa Forest Reserve, ii.1956, *Goldsmith* 87/56 (K; LISC; PRE; SRGH). C: Harare Distr., Lake McIlwaine, 16.xii.1965, *Plowes* 2548 (BR; K; LISC; PRE; SRGH). E: Nyanga Distr., Juliasdale, west of Punch Rock, c. 1925 m, 24.xii.1972, *Biegel* 4122 (K; LISC; MO; PRE; SRGH). **Malawi**. N: Mzimba Distr., 8 km W of S49 towards Vuvumwe Bridge, 20.i.1978, *Pawek* 13642 (BR; MO; SRGH; WAG). C: Lilongwe-Dzalanyama Road, near Katete Bridge, 6.ii.1957, *Robson* 1483 (BM; K; LISC; MAL; PRE; SRGH).

Also in Angola, Namibia, Zaire (Shaba) and Tanzania. In various grassland and woodland types usually in sandy soil, often in sandy dambos, also in submontane grassland; flowering mostly in December to March.

Easily recognised by its tufted, more or less stemless growth form and large, white flowers with perianth tubes 10–14 cm long.

The corms are part of the diet of the Khu Tribe of Namibia and Botswana.

9. DIERAMA K. Koch

Dierama K. Koch in Index Seminum Hort. Bot. Berol. **1854**, App.: 10 (1855). —Hilliard, Burtt & Batten, Dierama: Hairbells of Africa (1991).

Evergreen perennial herbs with large corms, aerial parts persisting for several years; corms with coarsely fibrous tunics. Leaves several, the lower 2–3 sheathing the stem base (cataphylls), these often dry and becoming fibrous; the foliage leaves linear, plane, with many strong parallel fibro-vascular veins, often without a midrib. Stem terete, slender and wiry, usually branched. Inflorescence consisting of spikes in a lax panicle; spikes few to many, erect or pendulous, terminal on the main axis and at the ends of wiry lateral branches of the flowering shoot; floral bracts ± solid, or scarious, and often translucent, lacerate above and usually brown-streaked or veined. Flowers usually pink (also red, purple, yellow or white in southern Africa), actinomorphic, usually pendent and campanulate, with a fairly short campanulate perianth tube; tepals subequal; stamens, and often the style, included. Stamens symmetrically disposed. Style exserted from the perianth tube, seldom exceeding the tepal lobes; style branches simple, short, filiform. Capsules globose, coriaceous. Seeds globose or lightly angled, hard, smooth and often shiny.

A genus of 44 species, extending from the E Cape Province of South Africa through east tropical Africa to Ethiopia. Thirty nine species occur in southern Africa and 8 in tropical Africa.

Recognised by the wiry peduncles, usually drooping spikes, the pendent actinomorphic flowers and the dry, either solid or scarious floral bracts, generally pale with brown streaks and veins. Considered to be closely allied to the southern African *Ixia*, it is placed in the *Ixioideae*, but is unusual in the subfamily in having an apparently primitive leaf anatomy, often including the absence of a leaf midrib [Rudall & Goldblatt in Bot. J. Linn. Soc. **106**: 329–345, 1991)].

1. Spikes erect; plants single-stemmed or forming small clumps - - - - 8. *plowesii*
- Spikes pendulous; plants single-stemmed or forming small to large clumps - - 2
2. Floral bracts with a solid patch of hard tissue at least in the lower part, this often drying to a uniform pale-buff or brown - - - - - - - - - - - - - 3
- Floral bracts without a solid patch, although the veins may be very closely set - - 4
3. Flowers 32–43 mm long, perianth tube 10–15 mm long; stigmas held 8–16 mm below the tepal apices - - - - - - - - - - - - - - - - - 1. *inyangense*
- Flowers 20–29 mm long, perianth tube 6.5–10 mm long; stigmas held 2.5–6.5 mm below the tepal apices - - - - - - - - - - - - - - - 2. *densiflorum*

4. Plants forming large clumps of several stems   -   -   -   -   -   -   -   -   5
- Plants solitary or forming clumps of few stems   -   -   -   -   -   -   -   -   7
5. Floral bracts rich rusty-brown with 6–8 veins each side of midvein fading out in upper third of the bract; leaves 2–4 mm wide   -   -   -   -   -   -   -   -   -   5. *pauciflorum*
- Floral bracts either not rich rusty-brown, or with fewer veins, or leaves broader   6
6. Floral bracts less than 2 internodes long; terminal spike 2–6-flowered   4. *cupuliflorum*
- Floral bracts usually at least 2 internodes in length; terminal spike 6–10-flowered
                                                                                3. *formosum*
7. Terminal spike 1–3(5)-flowered; floral bracts bright rusty-brown; stigmas reaching (6)9–15 mm below the tepal apices; leaves 2–4 mm wide   -   -   -   -   -   -   5. *pauciflorum*
- Combination of characters not as above (terminal spike with more flowers or floral bracts not bright rusty-brown or leaves broader)   -   -   -   -   -   -   -   -   -   8
8. Stigmas reaching to tepal apices or slightly exserted   -   -   -   -   -   -   -   9
- Stigmas reaching 4–13(25) mm below the tepal apices   -   -   -   -   -   10
9. Flowers 22–32 mm long; tepals 15–22 × 6–11 mm   -   -   -   -   6. *longistylum*
- Flowers 17–22 mm long; tepals 10–15 × 5–7 mm   -   -   -   -   -   7. *parviflorum*
10. Floral bracts more or less solidly coloured rusty-brown but at least heavily flecked; veins fading out in the upper third of the bract   -   -   -   -   -   -   -   4. *cupuliflorum*
- Floral bracts lighly coloured even when flecked all over; veins fading out about halfway up the bract   -   -   -   -   -   -   -   -   -   -   -   -   -   -   -   -   11
11. Floral bracts rarely up to 2 internodes long; stigmas reaching to 6 mm below the tepal apices
                                                                                6. *longistylum*
- Floral bracts reaching or exceeding 2 internodes in length; stigmas reaching (5)7–11 mm below the tepal apices   -   -   -   -   8. *plowesii variants*, see notes under this species.

1. **Dierama inyangense** Hilliard & Burtt in Notes Roy. Bot. Gard. Edinburgh **45**: 81 (1988). —Hilliard, Burtt & Batten, Dierama: Hairbells of Africa: 52 (1991). TAB. **11** fig. A. Type: Zimbabwe, Nyanga (Inyanga), 20.ii.1965, *Plowes* 2678 (SRGH, holotype; K).
    Dierama pendulum sensu Tredgold & Biegel, Rhod. Wild Fl.: 11, pl. 6 (1979) non *Dierama pendulum* (L.f.) Bak. (1877).

Plants either single-stemmed or growing in small tufts. Corms not seen. Leaves several, the basal ones 450–750 × 5.5–7.5 mm, sheathing leaves 4–8. Stems 120–200 mm long, with 3–6(or more) branches. Spikes pendulous, terminai spike 8–11-flowered, lateral spikes 6–11-flowered; flowers crowded and bracts imbricate, (the tip of each bract often reaching to the middle of the second bract above it); floral bracts 17–30 mm long, oblong-obovate to lanceolate, acute to acuminate, veins many, fusing into solid tissue covering half to nearly the whole of the bract; drying light-buff, or sometimes partly flecked, margins and shoulders whitish to heavily flecked or almost solidly brown. Flowers light to dark purple-pink, 32–43 mm long; perianth tube 10–15 mm long; tepals 22–28 × 8–10 mm. Anthers 8–10 mm long. Stigmas reaching to 8–16 mm below the tepal apices.

Zimbabwe. E: Nyanga (Inyanga), Pungwe Hills, 10.xi.1942, *Hopkins* in GHS 13707 (SRGH).
Mozambique. MS: Tsetserra, 1955, *Exell, Mendonça & Wild* 230 (SRGH).
Apparently endemic to the eastern border mountains of Zimbabwe and adjacent mountains of Mozambique, between Nyanga and Tsetserra, *Dierama inyangense* occurs in rocky grassland from c. 1650 to 2300 m; flowering between August and March.
It is closely allied to *D. insigne* N.E. Br. from the SE Transvaal, a generally smaller plant (leaves 2–4.5(6) mm wide, and flowers 22–36 mm long) with the stigma longer and usually reaching 1–8.5 mm below the tepal apices. In its broad leaves, it resembles the central Malawian and southern Tanzanian *D. densiflorum* Marais, which has solid bracts, but smaller flowers with longer stigmas that reach 2.5–6.5 mm below tepal apices.

2. **Dierama densiflorum** Marais in Notes Roy. Bot. Gard. Edinburgh **45**: 105 (1988). —Hilliard, Burtt & Batten, Dierama: Hairbells of Africa: 54 (1991). Type from Tanzania.
    Dierama pendulum sensu Plowes & Drummond, Wild Fl. Rhodesia: pl. 36 (1976) non *Dierama pendulum* (L.f.) Bak. (1877).

Plant single-stemmed or forming tufts of a few stems. Corms 20–30 mm in diameter. Leaves several, the basal ones 550–900 × 6.5–12 mm, sheathing cauline leaves 3–4. Stems 135–165 mm long, 2–5-branched, or more. Spikes pendulous, terminal spike 8–11-flowered, lateral spikes 4–8-flowered; flowers somewhat crowded (bracts usually c. 2 internodes long); floral bracts 16–23 mm long, obovate, acute; veins many, coalescing to form a solid patch reaching up to three-quarters of the way up the bract, drying light-buff, or sometimes flecked with brown, or rarely wholly brown, margins and shoulders heavily flecked. Flowers rose- to purple-pink, 20–29 mm long; perianth tube 6.5–10 mm long;

Tab. 11. A. —DIERAMA INYANGENSE, leaf and upper part of flower shoot (×½), dissected flower, style and bract (×1), from *Plowes* 2678. B. —DIERAMA PAUCIFLORUM, habit (×½), dissected flower, style and bract (×1), from *Wright* 1601. Drawn by A. Batten.

tepals 12–19 × 5–8.5 mm. Anthers 6–6.5(7) mm long. Stigmas reaching 2.5–6.5 mm below the tepal apices.

Malawi. N: Viphya, Mzimba River bridge, 14.iii.1962, *Chapman* 1625 (K; SRGH).
Also in SW Tanzania in the Mbeya and Poroto Mts. In open, often rocky submontane grassland at 1800–2100 m; flowering in March to July.

D. densiflorum is one of the small group of species with a distinctive type of floral bract in which the closely set veins coalesce to form a solid patch without intervening membranous tissue. The closely allied *D. insigne* N.E. Br. from the SE Transvaal has broader leaves and flowers in October and November.

3. **Dierama formosum** Hilliard in Notes Roy. Bot. Gard. Edinburgh **45**: 80 (1988). —Hilliard, Burtt & Batten, Dierama: Hairbells of Africa: 58 (1991). Type: Zimbabwe, summit of Inyangani Mt., c. 2600 m, 7.x.1962, *Plowes* 2268 (K, holotype; PRE).

Plants forming large tufts with many stems. Corms 30–35 mm in diameter. Leaves several, the basal ones 300–750 × (4)6–10 mm, sheathing cauline leaves 5. Stems 75–165 mm long, 3–5-branched. Spikes pendulous, the terminal spike 6–10-flowered, lateral spikes 3–9-flowered; flowers moderately crowded (bracts often 2 internodes long or longer); floral bracts 20–25 mm long, lanceolate to oblanceolate-obovate, acute to acuminate, solid red-brown or densely flecked, thus appearing solidly coloured, often slightly lighter towards the shoulders and margins where flecks are more dispersed, rarely entire bract merely flecked and then appearing lighter; main veins 5 on each side of the midvein, intermediate veins often well developed and then the veins very close and the base sometimes semi-solid, but veins always distinct, fading out in the middle or upper third of the bract. Flowers pale-lilac to bright cerise-pink, 26–37 mm long; perianth tube (6.5)8–12 mm long, tepals (17)19–26 × 6–11 mm. Anthers 7–9 mm long. Stigmas reaching 2.5–14 mm below the tepal apices.

Zimbabwe. E: Stapleford Forest Reserve, 8.x.1959, *Chase* 7173 (K; MO; PRE; SRGH). **Malawi.** S: Mt. Mulanje, Lichenya Plateau, 17.x.1941, *Greenway* 6317 (BR; K; PRE). **Mozambique.** MS: Gorongosa Mt., Gogogo summit, x.1971, *Tinley* 2208 (BR; K; PRE; SRGH).
Also in the eastern Transvaal (South Africa). Relatively common in montane grassland at 1370 to 2590 m; flowering in all months.

Allied to *D. cupuliflorum* Klatt, but more robust and often taller, and with larger corms and broader leaves. The more crowded terminal spikes of 6–10 flowers (3–9 in the lateral spikes) and distinctly elliptic tepals readily distinguish *D. formosum* from *D. cupuliflorum* which usually has 2–6 flowers per terminal spike (2–4 per lateral spike) and usually oblong tepals.

4. **Dierama cupuliflorum** Klatt in von der Decken, Reis. Ost-Afr., Bot. **3**: 73, t. 3 (1879). Type from Tanzania.

Subsp. **cupuliflorum** —Hilliard, Burtt & Batten, Dierama: Hairbells of Africa: 60 (1991). Type as above.
Dierama vagum N.E. Br. in J. Roy. Hort. Soc. **54**: 200 (1929). Type from Kenya.

Plants forming clumps of few to many stems. Corms 10–20(25) mm in diameter. Leaves several, the basal ones 200–850 × 3.5–7(11) mm, cauline sheathing leaves 5. Stems 30–135 mm long, 2–4-branched. Spikes pendulous, the terminal spike 2–6(7)-flowered, lateral spikes 2–4(5)-flowered; flowers usually laxly arranged (bracts usually less than 2 internodes long, or longer when the apical acumen is very well developed); floral bracts 14–26(32) mm long, oblong to obovate, acute to shortly acuminate, often more or less solidly coloured dark-brown, lighter but heavily flecked on the shoulders, or sometimes lighter in tone but heavily flecked; veins 5–6 on each side of midvein, up to 0.3 mm apart, but intermediate veins often strongly developed, fading out in upper third of bract. Flowers 18–30 mm long, pale-lilac to purple or reddish; perianth tube 4–8 mm long; tepals 12–15 × 4.5–9 mm. Anthers 5–7.5 mm long. Stigmas reaching to 3–6 below tepal apices.

Malawi. N: South Viphya plateau, 9.v.1970, *Brummitt* 10539 (K).
Also in Tanzania, Kenya, Uganda and Ethiopia. In montane grassland and heath between c. 2000 and 2200 m in northern Malawi; flowering almost any month.

Common on the East African mountains to the north. The oblong tepals, narrow in relation to their length are diagnostic for this species. The bracts are usually dark-brown, or sometimes light-brown, but always heavily flecked.

Subspecies *caudatum* Marais, restricted to SW Tanzania, is distinguished from subsp. *cupuliflorum* by its broader leaves up to 11 mm wide, its tailed floral bracts and its mostly larger flowers.

5. **Dierama pauciflorum** N.E. Br. in J. Roy. Hort. Soc. **54**: 200 (1929). —Hilliard, Burtt & Batten, Dierama: Hairbells of Africa: 64 (1991). TAB. **11** fig. B. Type from South Africa (Orange Free State).

Plant usually forming dense clumps, but occasionally stems solitary. Corms c. 15 mm in diameter. Leaves several, the basal ones 100–400 × 2–4 mm, sheathing cauline leaves usually 3. Stems 30–60 cm long, simple or with up to 5 branches. Spikes pendulous, the terminal spike 1–4(5)-flowered, lateral spikes 1–3(4)-flowered; flowers lax with bracts 1.5–2 internodes long; floral bracts 12–18 mm long, broadly obovate-oblong to broadly elliptic or lanceolate-elliptic, bright rust-brown and darkly flecked; main veins 6–8 on each side of the midvein, 0.25–0.5 mm apart, fading out in the upper third of the bract. Flower bright-pink to reddish (rarely white), 20–28 mm long; perianth tube 5–7 mm long; tepals 14–22 × 4.5–8.5 mm. Anthers 5–7 mm long. Stigmas reaching to (6)9–15 mm below the tepal apices.

Zimbabwe. E: Nyanga (Inyanga), 3.x.1966, *Grosvenor* 273 (K; SRGH).
Also in South Africa (Transvaal, Natal, Orange Free State) and Transkei. Apparently only in the Nyanga Highlands of Zimbabwe. On mountain slopes at 500 to 2400 m.
The slender plants with narrow leaves 2–4 mm wide, relatively short stems to 60 cm high, and spikes of 1–4(5) flowers distinguish the species.

6. **Dierama longistylum** Marais in Notes Roy. Bot. Gard. Edinburgh **45**: 105 (1988). —Hilliard, Burtt & Batten, Dierama: Hairbells of Africa: 66 (1991). TAB. **12** fig. A. Type from Tanzania.

Plants either single-stemmed or forming tufts of 2–3 stems. Corms 15–40 mm in diameter. Leaves several, the basal ones 250–700 × 2.5–5(7) mm, sheathing cauline leaves c. 5. Stems 55–120 cm long, (1)2–5-branched. Spikes pendulous, the terminal spike 3–7-flowered, lateral spikes 2–5-flowered; flowers usually lax (bracts rarely 2 internodes long); floral bracts 14–25 mm long, lanceolate, acute, thin-textured, sometimes wholly suffused light orange-brown, or the solid colour confined to the midline, flecked to margins, colour sometimes much paler and flecking lighter; main veins 5 on each side of the midvein, 0.25–0.3 mm apart, intermediate veins often strongly developed, fading out about halfway up the bract. Flowers pale to deep mauve-pink, 22–32 mm long; perianth tube 6.5–10 mm long; tepals 15–22 × 6–11 mm. Anthers 6–9.5 mm long. Stigmas either reaching or slightly exceeding the tepal apices or up to 6 mm short of the tepal apices.

Zambia. E: Nyika Plateau, c. 6 km S of Nganda Hill, 24.xi.1955, *Lees* 86 (BR; K). **Malawi**. N: Nyika Plateau, 26.x.1958, *Robson & Angus* 371 (K; PRE; SRGH).
Also in SW and central Tanzania. Mainly in montane grassland, 600 to 2400 m; flowering mainly in September to January.
Allied to *D. cupuliflorum* Klatt and distinguished from it by the longer stigma, which reaches up to 6 mm below the top of the tepals (reaching 13–25 mm below the tepal apices in *D. cupuliflorum* subsp. *caudatum*; 4–13 mm below in subsp. *cupuliflorum* or exserted). The colour of the floral bracts always distinguishes the two, being much lighter in *D. longistylum*, and thinner in texture with the veins fading out about halfway up the bract. In the field they are usually easily distinguished by a difference in habit, the stems are usually clumped in *D. cupuliflorum*, and more or less solitary in *D. longistylum*.

7. **Dierama parviflorum** Marais in Notes Roy. Bot. Gard. Edinburgh **45**: 105 (1988). —Hilliard, Burtt & Batten, Dierama: Hairbells of Africa: 68 (1991). Type from Tanzania.

Plants single-stemmed. Corms c. 15 mm in diameter. Leaves several, the basal ones often absent at flowering time (burnt remains of previous seasons growth often visible), c. 400–600 × 2–4.5(9) mm, sheathing cauline leaves c. 4–5. Stems 40–85 cm long, 2–5 branched. Spikes pendulous, the terminal spike 3–7-flowered, lateral spikes 2–7-flowered; flowers moderately crowded (bract almost to about 2 internodes long); floral bracts 12–20 mm long, lanceolate, acute, ground colour whitish to light red-brown, flecked with brown, sometimes solid brown at the extreme base and along the midline; main veins 5–6 on each side of the midvein, intermediate veins well developed and therefore nearly contiguous, fading out in upper third of the bract. Flowers light to dark-mauve or lilac-pink, 17–22 mm long; perianth tube (5)6–7 mm long; tepals 10–15 × 5–7 mm. Anthers 6–7 mm long. Stigmas reaching to tepal apices or shortly exceeding them.

Zambia. N: Lumi Marsh (Zambia-Tanzania border), 1.xi.1956, *Richards* 6828 (K).
Also in SW Tanzania and Zaire (Shaba). Restricted to the highlands around the southern end of

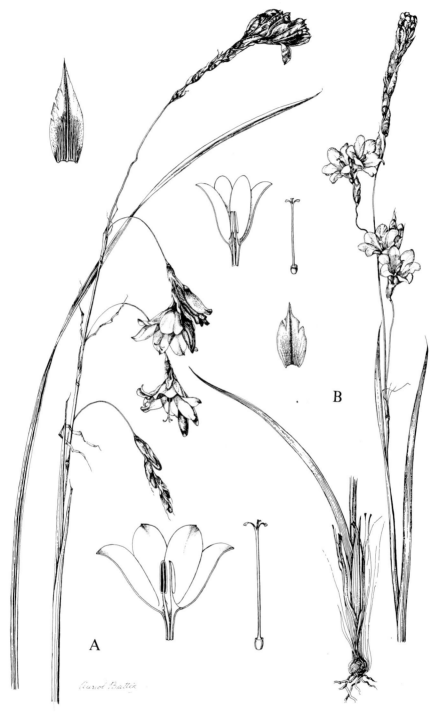

Tab. 12. A. —DIERAMA LONGISTYLUM, leaf and upper part of flower shoot (×⅓), dissected flower, style and bract (× 1), from *la Croix* 4182. B. —DIERAMA PLOWESII, whole plant (×⅓), dissected flower, style and bract (× 1), from *Plowes* 3150. Drawn by A. Batten.

Lake Tanganyika, between c. 1500 and 2450 m, and found in dambos or open grassland near woodland; flowering in September to November.

Closely allied to *D. longistylum* Marais which differs in having leaves present at the time of flowering and larger flowers (22–32 mm long).

8. **Dierama plowesii** Hilliard in Hilliard, Burtt & Batten, Dierama: Hairbells of Africa: 70, 143 (1991). TAB. **12** fig. B. Type: Zimbabwe, Chimanimani Mts., The Saddle, c. 1500 m, 10.xi.1968, *Plowes* 3150 (SRGH, holotype).

Dierama pendulum sensu Rendle in J. Linn. Soc., Bot. **40**: 210 (1911) pro parte, non *Dierama pendulum* (L.f.) Bak. (1877).

Plants single-stemmed. Corms 8 mm in diameter. Leaves several, the basal ones absent at flowering time (only the burnt bases of previous seasons growth present, these 3–4 mm wide); sheathing cauline leaves 6, these c. 250–400 × 3–4 mm. Stems 45–50 cm long, 2–3-branched. Spikes erect, the terminal one 4–6-flowered, lateral spikes 2–5-flowered; flowers moderately crowded (bracts up to 2 internodes long); floral bracts 13–15 mm long, elliptic-obovate, acute to shortly acuminate, white lightly flecked nearly all over, or more strongly flecked all over and then the midline solidly coloured red-brown; main veins 5 on each side of the midvein, fading out in the lower half of the bract, 0.3–0.4 mm apart but the intermediate veins often strongly developed. Flowers light magenta-pink, 21–22 mm long; perianth tube 5.5–6 mm long, tepals 15.5–16 × 5.5–6 mm. Anthers 6.5–7 mm long. Stigmas reaching 4–6 mm below the tepal apices.

Zimbabwe. E: Chimanimani Mts., The Saddle, c. 1500 m, 10.xi.1968, *Plowes* 3150 (SRGH). **Mozambique**. 'Portuguese South East Africa' (probably Chimanimani Mts.), 1907, *Alexander* s.n. (E).

Evidently endemic to the Chimanimani Mountains, on the Zimbabwe-Mozambique border, growing in open grassland at 1350–2200 m; mostly flowering in September to November.

Probably most closely related to *D. parviflorum* Marais from which it is distinguished by its usually broader floral bracts with veins fading in the lower half (not in the upper third), by its longer tepals (15.5–16 versus 10–15 mm) and by its included stigmas (not reaching or exceeding the tepal apices). The erect flowering spikes are an outstanding feature of *D. plowesii*, but the constancy of this character needs verification.

Dierama plowesii variants

Plants apparently intermediate between *D. plowesii* and *D. formosum* occur in the eastern highlands of Zimbabwe, around Mutare in the Stapleford Forest Reserve and near Chimanimani (Melsetter). These are often taller and have more flowers per spike than does *D. plowesii* and may have pendulous lateral spikes. The bracts range from white and lightly flecked red-brown, to entirely light red-brown in the same population (see Hilliard & Burtt, loc. cit. for detailed discussion). The two species require further field investigation.

10. TRITONIA Ker Gawl.

Tritonia Ker Gawl. in Bot. Mag. **16**: t. 581 (1802). —de Vos in J. S. African Bot. **48**: 105–163 (1982); tom. cit. **49**: 347–422 (1983).

Perennial herbs with globose corms, aerial parts dying back annually; corms with fibrous tunics. Leaves several to many, the lower 2 entirely stem sheathing (cataphylls); foliage leaves mostly basal, or the upper ones cauline, blades mostly plane (sometimes terete or H-shaped in section outside the Flora Zambesiaca area). Inflorescence an erect or inclined spike; floral bracts membranous to scarious, the outer often 3-dentate, the inner 2-dentate and shorter than, or about as long as, the outer. Flowers often orange, or yellow, pink or purple, actinomorphic or zygomorphic; perianth tube short to long, tepals subequal or bilabiate with the lower 3 often each bearing an adaxial tooth-like callus in the midline. Stamens usually unilateral and arcuate (symmetrical in one S African spp.). Style 3-branched, the branches filiform becoming somewhat broadened above. Capsules globose to ellipsoid, more or less membranous. Seeds numerous per locule, angular to globose.

A genus of 28 species, mostly in South Africa, with 2 species in the Flora Zambesiaca region.

Closely allied to *Crocosmia*, but this latter genus differs in its capsules which are globose to depressed, 3-lobed and more or less coriaceous, and in its seeds which are hard and shiny and few per locule.

Note: The specimen of *Tritonia nelsonii* Baker labelled "from Botswana, *Pole Evans & Ehrens* 1915", is in fact from near Buurmansdrift, W Transvaal, South Africa, and was probably collected by Ehrens alone. This species may grow in Botswana but has not yet been recorded from there.

Perianth tube 12–16 mm long, equalling or shorter than the tepals   -   -   -   1. *laxifolia*
Perianth tube c. 20 mm long, and 1.5–2 times as long as the tepals   -   -   -   - 2. *moggii*

1. **Tritonia laxifolia** (Klatt) Benth. ex Baker, Handb. Irid.: 195 (1892); in F.C. **6**: 126 (1896) excl. var. *strictifolia*. —de Vos in J. S. African Bot. **49**: 392 (1983). TAB. **13**. Types from South Africa (Cape Province).

    Montbretia laxifolia Klatt in Linnaea **32**: 754 (1863); in Abh. Naturf. Ges. Halle **15**: 359 (1882). —Stapf in Bot. Mag. **150**: t. 9038 (1928) (as *M. laxiflora*). Types as above.

    Tritonia bakeri Klatt in T. Durand & Schinz, Consp. Fl. Afric. **5**: 203 (1895) as nom. nov. pro *T. laxifolia* sensu Baker, nom. illegit. non Klatt (1882). Type from Tanzania.

    Tritonia clusiana Worsley in Gard. Chron. **38**: 269 (1905). Type from South Africa (Natal).

    Tritonia bracteata Worsley in Gard. Chron. **39**: 2 (1906). Type from South Africa (Natal).

Plants (10)20–60 cm high. Corms 10–15 mm in diameter, often with small basal cormlets; tunics of fine fibres. Foliage leaves 4–7, lanceolate, sometimes narrowly so, suberect or sometimes spreading; the lower ones basal and longer than the rest, usually reaching to about the base of the spike, 5–10(12) mm wide; upper leaves cauline and smaller. Stem erect or suberect, simple or 1–2-branched, rarely more branched. Spike fairly lax, secund, 6–12-flowered; outer bracts reddish-brown, often with fine brown stippling towards the apex, membranous with wide brownish papery margins, 6–12(18) mm long, 3(5)-toothed or subacute, striate, with the median vein stronger and ending in a dark, sometimes reduced and short middle tooth; inner bracts similar but 2-toothed and sometimes slightly shorter than the outer, the teeth dark-brown apically. Flowers orange to pale brick-red, zygomorphic, the lower 3 tepals each with a yellow or greenish median mark outlined in red and with a prominent tooth-like callus; perianth tube 12–16 mm long, oblique-campanulate, narrowly tubular below; tepals unequal, (10)12–15 mm long, obtuse, the upper 8–12 mm wide, broadly obovate, suberect, slightly hooded, the other tepals 5–8 mm wide and elliptic. Stamens unilateral and arcuate; filaments 12–15 mm long; anthers 4–6 mm long. Style dividing near the apex of the anthers, the branches 2–4 mm long. Capsules 12–15 mm long, ellipsoid-trigonous.

Zambia. N: Mbala Distr., above Kawimbe, 29.i.1957, *Richards* 8002 (K). E: Chipata (Fort Jameson), Kapatamoyo, 5.i.1959, *Robson* 1040 (K; SRGH). **Malawi**. N: Mzimba Distr., Vipya Plateau, E of Champhila (Champira) Mt., Lwanjati Peak, 11.i.1975, *Pawek* 8903 (K; MAL; MO). C: Dedza Distr., Ciwawo (Ciwau) Hill, 15.iii.1961, *Banda* 411 (MAL; SRGH). S: Zomba Mt., 16.i.1957, *Whellan* 1161 (MAL; SRGH).

Also in South Africa (E Cape Province, Natal) and Tanzania, but absent from Zimbabwe and Mozambique. The break in the distribution range seems correct. Plateau woodland and open grassland on mountain slopes; flowering in January to May (July).

2. **Tritonia moggii** Oberm. in Kirkia **3**: 24 (1963). —de Vos in J. S. African Bot. **49**: 401 (1983). Type: Mozambique, Inhaca Island, SW of Saco Bay, 5.vii.1957, *Mogg* 27315 (PRE, holotype; J; K; SRGH).

    Tritonia sp. cf. laxiflora —Mogg in Macnae & Kalk, Nat. Hist. Inhaca Isl., Moçamb.: 143 (1958).

Plants 20–50 cm high. Corms 10–15 mm in diameter, sometimes with small cormlets on stolons; tunics of fine fibres. Foliage leaves 6–9, mostly basal, 15–50 cm long, 4–8(11) mm wide, just exceeding the spike, linear-lanceolate, soft-textured; upper leaves cauline and smaller. Stem erect, simple or 1–3-branched. Spike lax, subsecund, usually 8–10-flowered; outer bracts brownish above, scarious, papery, 6–12(15) mm long, 2–3-toothed or acute, finely veined; inner bracts similar, sometimes slightly shorter, 2-toothed. Flowers orange or salmon-pink, zygomorphic, the 3 lower tepals each with a yellow basal mark outlined in red, the lowermost or all three with a small median callus; perianth tube (16)25–35(40) mm long, narrowly tubular, widened slightly near the apex; tepals ± unequal, (12)15–23 × 6–9 mm, oblong-lanceolate, the upper slightly wider erect and hooded. Stamens unilateral and arcuate; filaments 12–16 mm long; anthers 6–7 mm long, yellow. Style dividing near the anther apices, style branches c. 4 mm long. Capsules coriaceous, 8–15 mm long, ellipsoid or turbinate.

Tab. 13. TRITONIA LAXIFOLIA, whole plant (×⅔), from *la Croix* 290. Drawn by J.C. Manning.

Mozambique. GI: Inhambane, 18.vi.1958, *Torre* 1569 (COI). M: Inhaca Island, 5.iv.1944, *Mogg s.n.* (J, 28539; K); 2.viii.1984, *Zunguze & Boane* 769 (LMU; MO; WAG).

Restricted to southern Mozambique; low-lying sandy coastal plains; flowering in April to July (October).

Readily distinguished within *Tritonia* by the long perianth tube (usually 25–35 mm long) and usually only the lower median tepal with a callus. Plants with shorter tubes, 16–20 mm long (e.g. *Pedro & Pedrógão* 1331 (BR; COI; K; LMU; PRE) from Chibuto, Maniquenique) were thought by de Vos to belong in *T. moggii* but have perhaps acquired genes of *T. laxifolia* by hybrid introgression.

11. CROCOSMIA Planch.

Crocosmia Planch. in Fl. Serres Jard. Eur. **7**: 161 cum tab. (1851). —de Vos in J. S. African Bot. **50**: 463–502 (1984).

Perennial herbs with large corms, aerial parts dying back annually; corms with fibrous tunics. Leaves several, the lower 2–3 sheathing the stem base (cataphylls); foliage leaves plane or plicate, mostly basal, cauline leaves much reduced. Stem terete, usually branched. Inflorescence a spike, usually inclined or horizontal; floral bracts coriaceous, fairly short, the inner about as long as the outer. Flower colours in shades of orange to reddish, actinomorphic to strongly zygomorphic with stamens arched under the upper tepal, this sometimes much larger than the others, perianth tube cylindric, funnel-shaped or curved with a narrow cylindrical lower portion and a broad cylindrical upper portion. Style exserted, the branches filiform or apically broadened, entire or notched at the apex. Capsules globose to depressed and 3-lobed, more or less coriaceous; seeds globose, hard and shiny, few per locule.

A genus of 9 species, mostly in the southern Cape Province of South Africa, 2 species also occur in tropical Africa, and 1 in Madagascar. Closely related to *Tritonia*, the latter differing mainly in its ovoid capsules with several small seeds per locule.

The hybrid *C. x crocosmiiflora* (Lemoine) N.E. Br. is a common garden plant in southern African gardens, sometimes becoming naturalised, although not in the Flora Zambesiaca region. It has become a weed in Madagascar, eastern Australia, tropical Asia, and parts of tropical and N America. *C. x crocosmiiflora* can be distinguished from *C. aurea* by its zygomorphic flowers with unilateral stamens, although the tepals are subequal.

Flowers actinomorphic and pendent; tepals subequal and somewhat longer than the perianth tube - - - - - - - - - - - - - - - - - - 1. *aurea*
Flowers zygomorphic, facing to the side; tepals unequal with the upper (adaxial) longest, slightly shorter than the perianth tube - - - - - - - - - - 2. *paniculata*

1. **Crocosmia aurea** (Pappe ex Hook.) Planch. in Fl. Serres Jard. Eur. **7**: 161 cum tab. (1851). —Klatt in Peters, Naturw. Reise Mossambique: 516, t. 57 (1864); in T. Durand & Schinz, Consp. Fl. Afric. **5**: 203 (1895). —Baker in F.T.A. **7**: 355 (1898). —de Vos in J. S. African Bot. **50**: 476 (1984). —Rendle in J. Linn. Soc., Bot. **40**: 210 (1911). —Brenan in Mem. N.Y. Bot. Gard. **9**,1: 84 (1954). Type from South Africa (Cape Province).

Tritonia aurea Pappe ex Hook. in Bot. Mag. **73**: t. 4335 (1847). Type as above.
Babiana aurea (Pappe ex Hook.) Klotzsch in Allg. Gartenzeitung **19**: 293 (1851).
Crocanthus mossambicensis Klotzsch in Peters, Naturw. Reise Mossambique: 516 (1864) nom. inval.

Plants usually 60–100 cm high. Corms 10–20 mm in diameter, producing stolons up to 18 cm long; tunics membranous, almost papery, brown. Foliage leaves several, shorter or longer than the stems, linear-lanceolate, plane, thin-textured. Stem arching outward above, unbranched or (2)3–4-branched with the branches diverging. Spike 4–12-flowered, strongly arched, usually flexuous; bracts firm, green or becoming brown, 5–8(15) mm long, the outer acute to acuminate or 2–3-toothed, the inner 2-toothed. Flowers bright-orange (rarely with dark markings at the tepal bases), actinomorphic, hypocrateriform, usually directed toward the ground; perianth tube cylindric-curved, 10–16(22) mm long, barely expanded above; tepals subequal, 12–28 × 3–6 mm, narrowly elliptic, obtuse, spreading horizontally. Stamens symmetrically disposed; filaments 12–30 mm long, erect, exserted for most of their length, surrounding the style; anthers whitish to orange, 3–7 mm long. Style almost central, dividing opposite the anthers, branches 2–5 mm long. Capsules c. 10 mm long, 12–15 mm in diameter, depressed globose, bright orange-red on the inside. Seeds 1–2(4) per locule, blackish, glossy, 4–6 mm in diameter, globose, later becoming wrinkled.

Throughout eastern southern Africa, extending north across the equator into Uganda and Central African Republic. Distinctive in the genus in its actinomorphic flowers with the tube down-curved so that the flowers face the ground. Two subspecies are recognised: subsp. *aurea* is the more common and widespread; subsp. *pauciflora* is centred in western Zambia and Angola.

Stem normally 3–4-branched and often exceeding the leaves; perianth tube 13–27 mm long; tepals 14–35 mm long - - - - - - - - - - - - - - subsp. *aurea*
Stem unbranched and shorter than the leaves; perianth tube 10–12(15) mm long; tepals c. 12 mm long - - - - - - - - - - - - - - subsp. *pauciflora*

Subsp. **aurea**, TAB. **14** fig. A.

Plants 60–100 cm high. Stem strongly arched above, usually 3–4-branched. Spikes 7–12-flowered, flexuous, exceeding the leaves; floral bracts 4–6(9) mm long. Perianth tube 12–16(22) mm long; tepals 15–28 × 5–6 mm. Filaments 20–30 mm long; anthers 5–7 mm long. Style branches (3)4–5 mm long, usually exceeding the anthers.

Zambia. N: Kasama Distr., Chishimba Falls, 12.ii.1961 *Robinson* 4373 (K; SRGH). **Zimbabwe**. E: Himalayas, Engwa, 2.iii.1954, *Wild* 4445 (K; MO). **Malawi**. N: Nyika Plateau, edge of Juniper Forest, 8.vi.1981, *Salubeni & Tawakali* 3053 (MAL; MO; P; SRGH). C: Ntchisi Forest, 3.v.1961, *Chapman* 1248 (MAL). S: Mulanje, Great Ruo Gorge, 12.v.1967, *Patrick* s.n. (MAL). **Mozambique**. MS: between Chimoio and Garuso (Garozo), 24.iii.1948, *Mendonça* 3851 (LISC; MO). Z: Gurué, Cascata, Rio Nacabe, 1240 m, fr. 4.viii.1979, *de Koning* 7571 (MO; WAG). M: Maputo, Matutuíne (Bela Vista) between Zitundo and Ponto do Ouro, 8.iii.1968, *Gomes e Sousa & Balsinhas* 5055 (PRE).
Also in South Africa (Cape Province, Natal, Transvaal), Tanzania and Zaire. Mostly in forest margins and clearings in areas of high rainfall; flowering in (December) February to April, sometimes until June.

Subsp. **pauciflora** (Milne-Redh.) Goldblatt comb. nov.
    Tritonia cinnabarina Pax in Bot. Jahrb. Syst. **15**: 152 (1893). —Baker in F.T.A. **7**: 357 (1898). Type from Angola.
    Crocosmia pauciflora Milne-Redh. in Kew Bull. **3**: 469 (1948). Type: Zambia, Mwinilunga Distr., by Kaoomba R., 22.xii.1937, *Milne-Redhead* 3783 (K; PRE, lectotype selected by de Vos).
    Crocosmia cinnabarina (Pax) de Vos in J. S. African Bot. **49**: 415 (1983); **50**: 500 (1984).
    Crocosmia aurea var. *pauciflora* (Milne-Redh.) de Vos in J. S. African Bot. **50**: 482 (1984).

Plants 45–80 cm high. Stem suberect, rarely branched. Spike (3)4–7-flowered, barely flexuous, shorter than the leaves; bracts 2–4 mm long. Perianth tube 10–12 mm long; tepals 12–14 × 3–4.5 mm. Filaments 12–15 mm long; anthers 3–3.5 mm long. Style branches 2–2.5 mm long, not or barely exceeding the anthers.

Zambia. W: Mwinilunga Distr., c. 20 km SSE of Salujinga, road to Kalene Hill, 21.ii.1975, *Hooper & Townsend* 279 (C; K; NDO; SRGH).
Also in Angola and S Zaire. Forest margins and clearings, and streamsides; flowering in December to February.
Raised from varietal to subspecies rank here, this seems a true geographical race, easily separable from subsp. *aurea* by its smaller flowers less crowded and on nearly straight spikes that do not exceed the leaves. The possibility that this should be recognised as a distinct species remains unsettled. Plants from NE Zaire, Burundi and Central African Republic are similar in being unbranched and having small flowers, but are more robust and have longer floral bracts. They may represent another distinct race of *C. aurea* or a variant of subsp. *pauciflora*.

2. **Crocosmia paniculata** (Klatt) Goldblatt in J. S. African Bot. **37**: 444 (1971). —de Vos in J. S. African Bot. **50**: 490 (1984). TAB. **14** fig. B. Type from South Africa (Natal).
    Antholyza paniculata Klatt in Linnaea **35**: 379 (1865). —Baker in F.C. **6**: 168 (1896). Type as above.
    Curtonus paniculatus (Klatt) N.E. Br. in Trans. Roy. Soc. S. Africa **20**: 270 (1932).

Plants 100–130(180) cm high. Corms single or in chains of 2–3, 2.5–4(5) cm in diameter; tunics brown, membranous, later fibrous above. Foliage leaves several, basal and cauline, half to two-thirds as long as the stems, lanceolate, plicate, narrowed and thickened below into a pseudopetiolate base; cauline leaves shorter than the basal. Stem inclined above, relatively thick, up to 9 mm in diameter, 2–5-branched, the branches nearly horizontal. Spikes rather congested, 10–22-flowered, flexuous; bracts firm, green, becoming reddish-brown, 7–10 mm long, exceeding the internodes, the outer obtuse or apiculate, the inner

Tab. 14. A. —CROCOSMIA AUREA subsp. AUREA, whole plant (×⅔), from *Chapman & Chapman* 8422. B. —CROCOSMIA PANICULATA, upper part of flowering shoot, (×⅔), from *Goldblatt* 6860. Drawn by J.C. Manning.

shortly 2-toothed. Flowers orange to red or red-brown, zygomorphic, facing to the side; perianth tube curved, (25)30–40(45) mm long, tubular in the lower half, infundibuliform above; tepals unequal, obtuse, the uppermost longest, (10)15–18 × 6–9 mm, the upper laterals 7–10 mm long, spreading, the lower 3 tepals 6–12 mm long, the median one smallest. Stamens unilateral, arched under the uppermost tepal; filaments 25–40 mm long; anthers 6–8 mm long. Style dividing near the anther apices, the branches 4–6 mm long. Capsules 12–14 mm long, globose; seeds 4 mm in diameter, globose or angled.

Zimbabwe. E: Mutare Distr., Vumba, Cloudlands, 8.i.1957, *Chase* 6315 (SRGH).

Also in South Africa (Natal, Transvaal, Orange Free State), Lesotho and Swaziland. Moist rocky and grassy slopes, often at high elevations, sometimes at forest edges; flowering in December to February.

Easily distinguished from *C. aurea* by its long, rather tubular flower and the broad plicate leaves.

12. ANOMATHECA Ker Gawl.

Anomatheca Ker Gawl. in Ann. Bot. (König & Sims) **1**: 227 (1804). —Goldblatt in Contrib. Bolus Herb., No.4: 75–91 (1972).
Lapeirousia subgen. *Anomatheca* (Ker Gawl.) Baker, Handb. Irid.: 173 (1892); in F.C. **6**: 89 (1896); in F.T.A. **7**: 351 (1898).

Perennial herbs with corms, aerial parts dying back annually; corms with fibrous tunics. Leaves several, the lower 1–2, entirely sheathing (cataphylls); foliage leaves plane, firm or soft-textured. Inflorescence a spike, usually flexed at the base (flowers solitary on inflorescence branches in 2 South African spp.), floral bracts herbaceous to coriaceous, the inner shorter than the outer, often brown at the apex. Flowers zygomorphic, in shades of pink to red or whitish usually with contrasting darker markings on the lower 3 tepals; perianth tube well developed, either slender and cylindric throughout or narrowly campanulate; tepals subequal or the upper larger and erect and the 3 lower inclined to the horizontal. Stamens unilateral and arcuate, filaments usually mostly included; anthers parallel and contiguous. Ovary globose-oblong, style exserted; style branches filiform, each deeply divided and recurved. Capsules globose, coriaceous, usually papillate to rugose. Seeds few to many per locule, angular to oblong-globose.

A genus of 5 species, mostly in South Africa. Closely related to (and perhaps congeneric with) the South African genus *Freesia*, the two differing significantly only in the shape of the perianth tube.

Tepals broadly ovate to oblong, held at right angles to the perianth tube, tepals less than half the length of the tube   -   -   -   -   -   -   -   -   -   -   -   -   1. *laxa*
Tepals lanceolate, forming a cup, tepals more than half as long as the perianth tube
2. *grandiflora*

1. **Anomatheca laxa** (Thunb.) Goldblatt in J. S. African Bot. **37**: 442 (1971); in Contrib. Bolus Herb., No.4: 83, fig. 6G (1972). TAB. **15** fig. B. Type from South Africa (Cape Province).
Gladiolus laxus Thunb., Fl. Cap.: 50 (1823). Type as above.
Meristostigma laxum (Thunb.) A. Dietr., Sp. Pl. **2**: 597 (1833).
Lapeirousia laxa (Thunb.) N.E. Br. in J. Linn. Soc., Bot. **48**: 24 (1928).

Plants 20–35 cm high. Corms conical, round-based, about 1 cm in diameter, with tunics of fine reticulate fibres. Foliage leaves several, usually exceeding the spike, narrowly lanceolate, soft textured. Stem erect, usually unbranched. Spike nearly horizontal, secund, 2–6-flowered; outer bracts 6–8(13) mm long, green, becoming membranous above, often dark-brown apically, inner bracts slightly smaller than the outer, bifid. Flowers zygomorphic, pink to red, or pale-blue to white, with either darker red or blue-violet marks at the base of the lower 3 tepals; perianth tube erect, slender, widening slightly at the apex, 18–33 mm long; tepals subequal, 9–13 cm long, broadly ovate to oblong, held at right angles to the tube. Stamens unilateral and arcuate, exserted from the tube for 1.5–2 mm; anthers 3–4 mm long. Style branching between the base and middle of the anthers; style branches c. 2.5 mm long, deeply forked, often tangled among the anthers. Capsules 9–12 × 8–10 mm, nearly smooth to lightly papillose in the upper half. Seeds c. 4 per locule, 2–3 mm in diameter, round, dark red-brown, glossy.

Two subspecies are recognised, one coastal from southern Mozambique and coastal Natal (South

Africa) and the other mostly inland and widespread, from South Africa (eastern Cape Province) to Uganda. *Anomatheca laxa* subsp. *laxa* is widely cultivated and has become naturalised in southern United States of America (Florida), in Madeira, Mascarene Islands and Hong Kong.

Flowers pink to red with dark-red markings on the lower tepals; upper tepal c. 12 × 5 mm
  - - - - - - - - - - - - - - subsp. *laxa*
Flowers white to pale-bluish with violet to indigo markings on the lower tepals; upper tepal
  c. 9 × 4 mm - - - - - - - - - - - - - - subsp. *azurea*

Subsp. laxa

    Anomatheca cruenta Lindl. in Bot. Reg. **16**: t. 1369 (1830). Type from South Africa (Cape Province).
    Lapeirousia cruenta (Lindl.) Baker, Handb. Irid.: 173 (1892); in F.C. **6**: 96 (1896); in F.T.A. **7**: 354 (1898). —Mogg in Macnae & Kalk, Nat. Hist. Inhaca Isl., Moçamb.: 143 (1958).
    Freesia cruenta (Lindl.) Klatt in T. Durand & Schinz, Consp. Fl. Afric. **5**: 187 (1895).
    Lapeirousia graebneriana Harms in Bot. Jahrb. Syst. **28**: 366 (1901). Type from Tanzania.

Plants with pink to red flowers, the lower 3 tepals each with a dark-red blotch in the lower third. Perianth tube 18–25(33) mm long; tepals 10–12 mm long, elliptic, c. 4 mm wide. Filaments exserted 1.5–2 mm; anthers 3–4 mm long. Style dividing opposite the middle of the anthers; style branches c. 2.5 mm long, forked for c. 1 mm, and tangled in the anthers.

Zambia. N: Saisi Valley, ? State (Start) Ranch, 20.v.1968, *Sanane* 145 (B; K). W: Mufulira, Kafue R., 27.ii.1960, *Fanshawe* 5379 (BR; K; NDO). C: Mpika Distr., Mpika-Serenje road, c. 1200 m, 5.iv.1961, *Richards* 15005 (K; MO; SRGH). **Malawi**. N: Viphya Plateau, 55 km SW of Mzuzu, 15.iii.1975, *Pawek* 9161 (K; MAL; MO; SRGH). **Mozambique**. M: Namaacha, M'Ponduine Monte (Mt. Ponduine), c. 800 m, 20.xii.1978, *Schäfer* 6640 (LMU; WAG).

Also in South Africa (Cape Province and Natal), Zaire (Shaba), S Tanzania, W Kenya and Uganda. On rocky hillsides and mountain slopes, especially cliffs, sometimes along forest margins; mostly flowering in November to January.

Subsp. azurea Goldblatt & Hutchings in Novon **3**: 146 (1993). Type: Mozambique, Maputo, Inhaca Island, 20.vii.1980, *de Koning & Navunga* 8312 (LMU, holotype; BR; K; MO; SRGH).

Plants with white to pale-bluish flowers, the lower 3 tepals each with a dark-violet to indigo blotch in the lower third. Perianth tube (25)27–32 mm long; tepals 9–11 mm long, elliptic, c. 4 mm wide. Filaments exserted for c. 1 mm; anthers c. 4 mm long. Style dividing just above the base of the anthers; style branches c. 2.5 mm long, forked for c. 1 mm, and tangled in the anthers.

Mozambique. GI: Massinga Distr., Rio das Pedras, 8.vii.1981, *de Koning & Hiemstra* 8933 (BR; LMU; MO; SRGH). M: Inhaca Island, 20.vii.1980, *de Koning & Navunga* 8312 (BR; K; LMU; MO; SRGH).
Also in South Africa (Natal). Along the coast in coarse sandy soil, usually in shade; mostly flowering in June to August.
Differing from subsp. *laxa* in the white to pale-blue flowers with dark-blue to violet markings at the base of the lower tepals, in the typically longer perianth tube which is usually 27–32 mm long and sometimes in the smaller tepals. Subspecies *laxa* with pink to red flowers has red marks at the bases of the lower tepals, a shorter tube usually 18–24 mm long, and somewhat larger tepals.

2. **Anomatheca grandiflora** Baker in J. Bot. **14**: 337 (1876). —Goldblatt in J. S. African Bot. **37**: 443 (1971). —Tredgold & Biegel, Rhod. Wild Fl.: 11, plate 7 (1979). —Plowes & Drummond, Wild Fl. Rhodesia: pl. 41 (1976). TAB. **15** fig. A. Type: Mozambique, mouth of Luaua R. (Luabo R.), 30.v.1858, *Kirk* s.n. (K, lectotype here designated).
    Lapeirousia grandiflora (Baker) Baker in Bot. Mag. **113**: t. 6924 (1887); Handb. Irid.: 173 (1892); in F.C. **6**: 96 (1896); in F.T.A. **7**: 355 (1898). —Schinz & Junod in Mém. Herb. Boissier, No. **10**: 30 (1900). —Eyles in Trans. Roy. Soc. S. Africa **5**: 332 (1916). —Suessenguth & Merxmüller, Contrib. Fl. Marandellas Distr.: 76 (1951). —Brenan in Mem. N.Y. Bot. Gard. **9**, 1: 83 (1954). —Letty, Wild Fl. Transvaal: 76, pl. 36 (1962).
    Freesia grandiflora (Baker) Klatt in T. Durand & Schinz, Consp. Fl. Afric. **5**: 187 (1895).
    Freesia rubella Baker in Bull. Herb. Boissier, sér. 2, **1**: 868 (1901). Type: Mozambique, Maputo (Delagoa Bay), 1890, *Junod* 166 (Z, holotype; K, flowers).
    Lapeirousia cuenta sensu Eyles in Trans. Roy. Soc. S. Africa **5**: 332 (1916) non *Lapeirousia cruenta* (Lindley) Baker (1892).
    Schizostylis coccinea sensu Mogg in Macnae & Kalk, Nat. Hist. Inhaca Isl., Moçamb.: 11, 143 (1958).

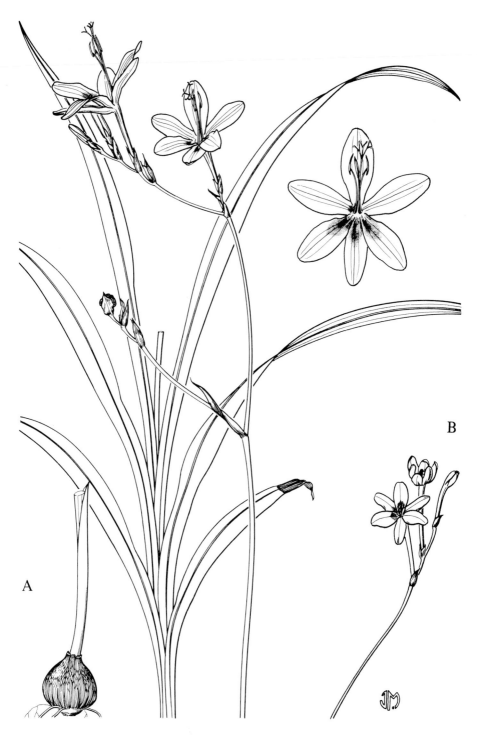

Tab. 15. A. —ANOMATHECA GRANDIFLORA, whole plant (×⅔), front view of flower (× 1), from *Snijman* s.n. B. —ANOMATHECA LAXA, flower spike (×⅔), from *Goldblatt* 9043. Drawn by J.C. Manning.

Plants (12)20–60 cm high. Corms 1–1.4 cm in diameter, globose; tunics of fine pale reticulate fibres. Foliage leaves several, usually reaching to the base of the spike but sometimes exceeding it, (6)8–12 mm wide, narrowly lanceolate. Stem erect, unbranched or 2–4-branched. Spike 2–6-flowered; outer bracts 10–15 mm long, green, becoming membranous to dry above, sometimes brown apically; inner bracts smaller than the outer, bifid. Flowers zygomorphic, red (pink), the lower 3 tepals each with a darker-red mark at the base; perianth tube erect, 20–30 mm long, campanulate, slender below, broad and cup-shaped above, the upper part, c. 4 mm long; tepals unequal, lanceolate, the upper 22–30(35) mm long, erect, the lower three 21–26 mm long, ascending to nearly horizontal. Stamens unilateral and arcuate, 15–25 mm long, exserted for (10)15–20 mm; anthers 5.5–8 mm long. Style branching between the middle and apex of the anthers, style branches 4–6 mm long, deeply forked, arched over the anthers. Capsules 8–10 mm long, 7–8 mm wide, smooth to irregularly papillate. Seeds nearly globose, c. 3 mm long, orange to dark-red, glossy.

Zambia. C: Lusaka, Church Road, 19.i.1986, *Goldblatt* 7576 (MO). E: Chipata (Fort Jameson), Mangwe Mt., 18.i.1962, *Grout* 272 (K; NDO). S: 35 km NE of Choma, 3.i.1957, *Robinson* 2007 (K; SRGH). **Zimbabwe**. N: Guruve (Sipolilo), Nyamunyeche (Nyamyetsi) Estate, 23.ii.1979, *Nyariri* 772 (MO; SRGH). W: Matobo Distr., Farm Besna Kobila, i.1955, *Miller* 2652 (K; LISC; SRGH). C: Esigodini (Essexvale), ii.1921, *Borle* 117 (K; SRGH; Z). E: Odzani R. Valley, 1915, *Teague* 338 (K). S: Nyoni Mts., 80 km S of Masvingo, 10.iii.1976, *Pope & Müller* 1520 (SRGH). **Malawi**. N: Mzimba Distr., Viphya, 12.viii.1966, *Agnew* 427 (MAL). C: Chongoni Mt., c. 1800 m, 3.ii.1959, *Robson* 1428 (K; LISC; MAL; SRGH). S: Blantyre Distr., south-facing slopes NW of Ndirande summit, 3.iii.1970, *Brummitt* 8853 (BR; K; MAL; SRGH). **Mozambique**. N: 7 km from Marrupa to Lichinga, 19.ii.1982, *Jansen & Boane* 7879 (K; LMU; MO). Z: Milange, Serra do Chiperone, 24.i.1965, *Correia & Marques* 2282 (BR; LMU; MO; WAG). T: Cahora (Cabora) Bassa, planalto do Songo, 850 m, 21.i.1973, *Torre, Carvalho & Ladeira* 18812 (LISC). M: Matutuíne (Bela Vista), 7.xii.1961, *de Lemos & Balsinhas* 252 (K; LMA; SRGH). MS: Inhamitanga Forest, 6.iv.1945, *Simão* 339 (LMA).

Also in South Africa (Natal, Transvaal), Swaziland, S Tanzania, and Zaire (Shaba). In light shade in deciduous woodland and scrub, coastal and inland, sometimes on anthills; mostly flowering in January to April.

Distinguished from *Anomatheca laxa* by its larger, usually entirely red flower with a wide upper perianth tube, erect upper tepals and long filaments exserted for at least 10 mm from the tube. Plants from near Mocuba in central Mozambique (*Barbosa & Carvalho* 2744) seem to represent an unusual variant of the species; the flowers are pink with reddish markings and the filaments are unusually short, c. 12 mm long, and included in the upper part of the tube.

13. BABIANA Ker Gawl.

Babiana Ker Gawl. in Bot. Mag. **16**: t. 576 (1802). —G.J. Lewis in J. S. African Bot., Suppl. 3 (1959).

Perennial herbs with globose corms, aerial parts dying back annually; corms with fibrous tunics. Leaves several, the lower 2–3 basal and entirely sheathing (cataphylls); foliage leaves usually plicate (or flat and sinuate or terete), usually pubescent or puberulent. Stem simple or branched, sometimes subterranean, often puberulent. Inflorescence a spike; floral bracts green, or dry and brown at least apically, the inner shorter than the outer and forked, sometimes cleft to the base. Flowers often in shades of blue or purple (sometimes yellow, red, pink or whitish in South African species), actinomorphic, or zygomorphic with stamens unilateral and arcuate; tepals united in a short to long perianth tube, subequal or forming 2 lips, sometimes the upper tepal much enlarged and arching over the stamens (or reflexed and sheathing the filaments). Ovary oblong-globose, sometimes villous, style exserted or included in the perianth tube; style branches short, somewhat expanded above. Capsules oblong-ovoid, coriaceous, often showing the outline of the seeds. Seeds blackish or dark-brown, several per locule, globose.

A taxonomically isolated genus of c. 62 species, mostly of dry areas in southern Africa. Best developed along the west coast of southern Africa. Also in S Zambia, Zimbabwe, Socotra and Namibia with one species widespread in south tropical Africa.

Babiana hypogea Burch., Trav. S. Africa **2**: 589 (1824). —Baker, Handb. Irid.: 180 (1892); in F.C. **6**: 107 (1896). —Dyer in Fl. Pl. Africa **25**: t. 962 (1946). —Martineau, Rhod. Wild Fl.: 18, pl. 2 (1954).

Tab. 16. BABIANA HYPOGEA, habit (×$\frac{2}{3}$), from *Biegel* 2075. Drawn by J.C. Manning

—Lewis in J. S. African Bot., Suppl. 3: 112 (1959). —Plowes & Drummond, Wild. Fl. Rhodesia: pl. 35 (1976). —Tredgold & Biegel, Rhod. Wild Fl.: 11, pl. 7 (1979). TAB. **16**. Type from South Africa (N Cape Province).

Antholyza hypogea (Burch.) Klatt in Abh. Naturf. Ges. Halle **15**: 345 (Erganz.: 11) (1882).
Babiana bainesii Baker in J. Bot. **14**: 335 (1876); in F.C. **6**: 107 (1896). —Eyles in Trans. Roy. Soc. S. Africa **5**: 331 (1916). —Suessenguth & Merxmüller, Contrib. Fl. Marandellas Distr.: 77 (1951). Type from South Africa (Transvaal).
Babiana schlechteri Baker in Bull. Herb. Boissier, sér. 2, **4**: 1005 (1904). Type from South Africa (Transvaal).
Babiana bakeri Schinz in Bull. Herb. Boissier, sér. 2, **6**: 712 (1906) as nom. nov. pro *B. schlechteri* Baker nom. illegit. Type as for *B. schlechteri* Baker.

Plants 12–20 cm high, sometimes forming clumps. Corms 2–3 cm in diameter, globose; tunics thickly matted, coarsely fibrous and tough, with a neck 6–20 cm long. Foliage leaves usually 4–7, erect or suberect, overtopping the flowers, 7–20(30) cm long, 3–10 mm wide, linear or lanceolate, laxly to deeply plicate, glabrous, hispidulous or pilose on the ribs. Stem subterranean, simple or branched below ground. Spike congested, 2–8-flowered, arising below ground; outer bracts green, 2.5–4(8) cm long, lanceolate, acuminate, scarious and ferruginous apically, inner bracts similar but slightly shorter and bifid at the apex. Flowers usually pale to dark-blue to mauve (rarely white), with a strong sweet spicy fragrance, zygomorphic, the lower lateral tepals each with a whitish streak in the midline; perianth tube 35–55 mm long, narrowly infundibuliform; tepals lanceolate, acute, the outer 3 or sometimes all tepals mucronate, 25–60 × 6–12(15) mm, the 3 upper slightly longer than the lower ones. Stamens unilateral and arcuate, 10–15 mm long; anthers 8–10 mm long. Style branching at or beyond the anther apices, the branches 5–7 mm long. Capsules globose-oblong.

Botswana. SW: c. 4 km NE of Tshobokwane Borehole, 14.v.1976, *Skarpe* S-70 (K). SE: Gaborone, University Campus, c. 950 m, 23.iv.1975, *Mott* 274 (UCBG). **Zambia**. S: Livingstone, 22.iii.1961, *Fanshawe* 6448 (NDO). **Zimbabwe**. N: Guruve (Sipolilo), Nyamunyeche Estate, 31.i.1979, *Nyariri* 670 (MO; SRGH). W: Matobo, Farm Besna Kobila, iii.1953, *Miller* 1716 (MO; SRGH). C: 12 km SSE of Gweru (Gwelo), 19.iv.1967, *Biegel* 2075 (MO; SRGH). E: Nyanga Distr., 1 km from Nyanga village to Juliasdale, 8.iii.1981, *Philcox et al.* 8935 (K; SRGH). S: c. 30 km N of Masvingo (Fort Victoria), 4.v.1962, *Drummond* 7949 (K; SRGH).

Also in Namibia and South Africa (Cape Province, Transvaal, Orange Free State). Usually in Kalahari Sand or stony laterite in open woodland or grassland; flowering in December to May.

14. ZYGOTRITONIA Mildbr.

Zygotritonia Mildbr. in Bot. Jahrb. Syst. **58**: 230 (1923). —Goldblatt in Bull. Mus. Natl. Hist. Nat., sér. 4, B, Adansonia **11**: 201 (1989).

Perennial herbs with globose corms, aerial parts dying back annually; corms with membranous to fibrous tunics. Leaves few, the lower 2–3 membranous and sheathing the stem base (cataphylls); foliage leaves contemporaneous with flowers, 2–4, more or less plicate or folded, usually with at least two major veins and lacking a distinct midrib, the lowermost larger than the others, the upper laminate and more or less entirely stem sheathing (or leaves of the flowering shoot elaminate and sheathing, with laminate foliage leaves then sometimes produced on separate shoots after flowering). Stem terete, often with several diverging branches. Inflorescence spikes with many flowers per axis; floral bracts firm-textured, short, dry at the apex only or becoming dry and rust-coloured throughout, the outer bract acute, the inner bilobed, slightly shorter to slightly longer than the inner. Flowers zygomorphic, relatively small; tepals united in a short tube, unequal, linear-spathulate, obtuse, the upper tepal much exceeding the others and arched over the stamens, the lower 3 smallest and channelled. Stamens unilateral and arcuate. Style arching over the stamens, undivided, stigmatic at the apex. Capsule globose-trigonous (but often only with 2 or 1 locule developed). Seeds globose to ellipsoid, glossy, with an obscurely reticulate surface, 1 or rarely 2 per locule.

A genus of 4 species extending from Senegal to western Tanzania, with 1 species occurring in N Malawi and Zambia.

Zygotritonia nyassana Mildbr. in Bot. Jahrb. Syst. **58**: 231 (1923). —Goldblatt in Bull. Mus. Natl. Hist. Nat., sér. 4, B, Adansonia **11**: 202 (1989). TAB. **17**. Type from Tanzania.

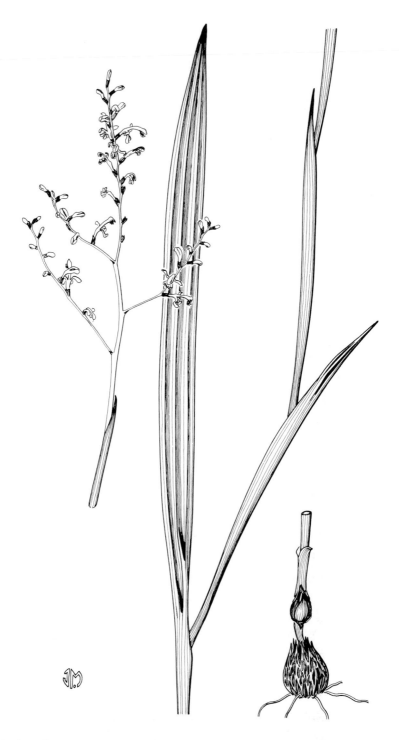

Tab. 17. ZYGOTRITONIA NYASSANA, whole plant (×$\frac{2}{3}$), from *Pawek* 12507. Drawn by J.C. Manning.

Zygotritonia gracillima Mildbr. in Bot. Jahrb. Syst. **58**: 232 (1923). Type from Zaire (Shaba).
Zygotritonia giorgii De Wild., Contrib. Fl. Katanga, Suppl. 1: 5 (1927). Type from Zaire (Shaba).
Zygotritonia homblei De Wild., Contrib. Fl. Katanga, Suppl. 1: 7 (1927). Type from Zaire (Shaba).

Plants 20–45 cm high. Corms 15–20 cm in diameter; tunics of moderately coarse reticulate to thick fibres, the fibres vertically thickened below and claw-like. Foliage leaves 3–4, lanceolate to nearly linear, 4–8(14) mm wide, plicate to weakly folded, with a single main vein or sometimes 2–3 primary veins; the lowermost 1 or 2 leaves basal or nearly so with the lowermost larger than the others, reaching at least to the base of the spike, sometimes just exceeding it; the upper leaves cauline and reduced in size. Stem (1)3–4-branched, the branches ascending. Spike with 20 or more flowers on the main axis; floral bracts green below and dry apically, densely papillate, the outer c. 2 mm long, the inner nearly 3 mm long. Flowers cream to greenish, reddish or purple; perianth tube 3–4 mm long, widening gradually from base to apex; tepals more or less linear-spathulate, the upper 10–12 mm long, slightly wider in the upper third and then about 2 mm wide, the other tepals 5–6 mm long, recurving and twisted loosely. Filaments c. 10 mm long; anthers c. 2 mm long. Style reaching to about the middle of the anthers, curving downward when receptive. Capsules 3–4 mm long.

Zambia. N: Isoka Distr., c. 18 km from Nakonde (Tunduma) to Mbala, 10.i.1975, *Brummitt & Polhill* 13691 (BR; K; NDO; SRGH; WAG). W: Kitwe, 10.ii.1954, *Fanshawe* 800 (K; SRGH); Nchanga, i.1942, *Ferrar* in Rhod. Mus. Herb. 4807 (K). **Malawi**. N: Chitipa Distr., Kaseye Mission, c. 16 km E of Chitipa, c. 1300 m, 18.iv.1976, *Pawek* 11085 (K; MAL; MO; SRGH).
Also in SW Tanzania and S Zaire (Shaba). In light woodland and open grassland; flowering in mid-December to March.
Closely allied to *Zygotritonia bongensis* which is largely West African in distribution, but is distinguished from it by having relatively small flowers and short bracts 2–3 mm long, the outer bracts being consistently smaller than the inner. The upper cataphyll is typically truncate with revolute margins.

15. HESPERANTHA Ker Gawl.

Hesperantha Ker Gawl. in Ann. Bot. (Konig & Sims) **1**: 224 (1804). —Goldblatt in Ann. Missouri Bot. Gard. **73**: 135 (1986).

Perennial herbs with small corms, aerial parts dying back annually; corms with woody to coriaceous tunics. Leaves few to several, the lower 2–3 membranous and entirely sheathing (cataphylls); foliage leaves lanceolate to linear (rarely terete), the blades plane or sometimes with raised margins and midrib. Stem simple or occasionally branched. Inflorescence a spike; floral bracts green, or membranous to dry apically, the inner smaller than the outer and bicarinate. Flowers usually white or pink, usually closed in the day and opening in the evening; actinomorphic, hypocrateriform (zygomorphic in one South African species); tepals united in a straight or curved perianth tube, subequal, patent-spreading or cup-shaped. Stamens usually symmetrically arranged; filaments straight (included within the perianth in a few South African species and unilateral in one); anthers facing inwards or articulated on the filaments and horizontal. Style usually dividing at the mouth of the perianth tube (or within the tube in a few South African species); style branches long and spreading, stigmatic along the entire length. Capsules broadly ovoid to cylindric, sometimes dehiscing only in the upper third. Seeds many, subglobose to angular, sometimes lightly winged on the angles.

A genus of c. 65 species, 63 in southern Africa, 3 in tropical Africa, 1 extending from eastern Zimbabwe through Tanzania to Ethiopia and the Cameroons.

1. Flowers borne erect, with a straight perianth tube; outer floral bracts with the margins free to the base - - - - - - - - - - - - - - 1. *petitiana*
– Flowers directed to one side of the spike and patent, or perianth tube curved to face the ground; outer floral bracts with margins united for 1–3 mm about the stem - - - - 2
2. Perianth tube 11–15 mm long, slightly longer than the bracts; spikes 1–2-flowered 3. *ballii*
– Perianth tube 18–30 mm long; spikes 3–6(9)-flowered - - - - - 2. *longicollis*

1. **Hesperantha petitiana** (A. Rich.) Baker in J. Linn. Soc., Bot. **16**: 96 (1878); in F.T.A. **7**: 348 (1898). —Goldblatt in Ann. Missouri Bot. Gard. **73**: 135 (1986). TAB. **18** fig. A. Type from Ethiopia.

Ixia petitiana A. Rich., Tent. Fl. Abyss. **2**: 309 (1850). Type as above.
Geissorhiza alpina Hook.f. in J. Linn. Soc., Bot. **7**: 223 (1864). Type from Cameroon.
Hesperantha alpina (Hook.f.) Pax ex Engl., Hochgebirgsfl. Afrika: 174 (1892). —Baker in
F.T.A. **7**: 348 (1898). —Hepper in F.W.T.A. ed. 2, **3**(1): 141 (1968).
Hesperantha volkensii Harms in Bot. Jahrb. Syst. **19**, Beibl. 47: 28 (1894). —Baker in F.T.A. **7**:
349 (1898). Types from Tanzania.
Hesperantha petitiana var. *volkensii* (Harms) R.C. Foster in Contrib. Gray Herb., No. 166: 22
(1948). —Brenan in Mem. N.Y. Bot. Gard. **9**(1): 84 (1954).

Plants 8–30(45) cm high. Corms 7–12 mm in diameter, globose to ovoid; tunics
dark-brown, woody to somewhat membranous, concentric. Leaves 3–4, the lower 2(3)
basal and longer than the others, 2–6 mm wide and about half to two-thirds as long as the
stem, the margins and midrib usually lightly thickened; upper leaf inserted in the middle
of the stem, short and usually entirely sheathing. Stem erect, very occasionally branched.
Spike (1)2–6(9)-flowered; floral bracts (9)12–15 mm long. Flowers pink, lilac or white;
perianth tube 6–9 mm long, cylindric and straight. Tepals (9)12–18 × 5–7 mm, ovate-
elliptic, the outer often flushed darker on the reverse. Anthers (3)4–7.5 mm long. Style
branches 6 mm long. Capsules 10–15 mm long, oblong-ellipsoid.

Zimbabwe. E: Mutare Distr., Engwa, 2250 m, 3.iii.1954, *Wild* 4473 (K; SRGH). **Malawi**. N: Nyika
Plateau, 14.viii.1946, *Brass* 17230 (K; MO; SRGH; US). C: Dedza Mountain, 3.iii.1977, *Grosvenor &*
Renz 1034 (K; MO; SRGH). S: Mulanje Mt., Chambe Plateau, 23.iii.1958, *Jackson* 2173 (K; SRGH).
Also in Tanzania, Kenya, Uganda, Ethiopia and Cameroon. On mountains above 1800 m, often in
rocky sites, on cliffs or in low grassland; flowering mostly in February to April in the Flora
Zambesiaca area.
H. petitiana is remarkably variable, sometimes only 8–12 cm high with 1–3 flowers per spike, but up
to 30 cm high and with six or more flowers per spike. It is difficult to distinguish from the South
African species *H. baurii* Baker, which in subsp. *baurii* usually has flexuous spikes of several flowers,
sometimes up to 10 per spike (the large flowered subsp. *formosa* from the Natal Drakensberg has
nearly straight spikes of 2–6 flowers). *H. baurii* has relatively short capsules 7–8 mm long, enclosed in
the bracts and usually no longer than the inner bract. In *H. petitiana* the capsules are 10–15 mm long,
at least as long as the inner bracts and often reaching or exceeding the outer bract apices.

2. **Hesperantha longicollis** Baker in Bull. Herb. Boissier, sér. 2, **4**: 1004 (1904). —R.C. Foster in
Contrib. Gray Herb., No. 166: 18 (1948). —Oberm. in Fl. Pl. Afr. **46**: pl. 1810 (1980) excl. syn. *H.*
sabiensis N.E. Br. ex R.C. Foster. —Goldblatt in Ann. Missouri Bot. Gard. **73**: 134 (1984). Type
from South Africa (Transvaal).
Hesperantha widmeri Beauverd in Bull. Herb. Boissier, sér. 2, **5**: 990 (1905). —R.C. Foster in
Contrib. Gray Herb., No. 166: 29 (1948). Type from South Africa (Transvaal).
Hesperantha matopensis Gibbs in J. Linn. Soc., Bot. **37**: 471 (1906). —Eyles in Trans. Roy. Soc. S.
Africa **5**: 330 (1916). Type: Zimbabwe, Matopos Hills, banks of the Malami R., ix.1905, *Gibbs* 44
(BM, holotype; BOL; K).

Plants 30–50(70) cm high. Corms c. 10 mm in diameter, campanulate; tunics dark-
brown, irregularly serrate on the basal margin, the outer layer often irregularly broken.
Foliage leaves 5–6, the lower 2–3 basal, from half to about as long as the stem, erect or
trailing, c. 2 mm wide, linear, the upper 1–2 leaves cauline, shorter than the basal ones
and partly to entirely sheathing. Stem erect, rarely with a branch from the base. Spike
3–6(9)-flowered, secund; floral bracts green to membranous with darker veins, 12–25 mm
long, the outer with the margins united around the stem for 1–2 mm, inner bract shorter
or longer than the outer. Flowers down-curved, whitish to ivory or light-brown, usually
darker on the reverse of the outer tepals; perianth tube 18–25 mm long, much exceeding
the bracts; tepals 14–18 × 5–6 mm, lanceolate, the inner slightly smaller than the outer.
Anthers yellow to brown, (6)8–10 mm long, pendent. Style branches 10–12 mm long,
exceeding the anthers. Capsules 10–12 mm long, narrowly ovoid to oblong, enclosed
within the bracts, dehiscing in the upper third.

Botswana. SE: Aedume Park, wet soil around Gaborone Dam, 4.viii.1978, *Hansen* 3422 (C; K; PRE;
SRGH). **Zambia**. B: Sesheke, without date, *Gairdner* 68 (K). **Zimbabwe**. W: Matobo Distr., farm
Besna Kobila, viii.1954, *Miller* 2450 (SRGH). C: Harare, Hatfield, 17.viii.1958, *Whellan* 1555 (BR; K;
PRE; SRGH). **Malawi**. N: 8 km NW of Lake Kaulime, 16.v.1970, *Brummitt* 10829 (K).
Also in South Africa (Transvaal). Along streams and in vleis and seeps; flowering mostly in August
and September, before the rains begin; flowers opening c. 5 pm.
Related to the South African species *H. radiata* (Jacq.) Ker Gawl., differing in the longer perianth
tube, narrow leaves and the outer bracts which are united for a shorter distance around the spike
axis.

Tab. 18. A. —HESPERANTHA PETITIANA, whole plant (×⅔), flower (×1), from *Phillips* 1731.
B. —SCHIZOSTYLIS COCCINEA, whole plant (×⅔), from *Davies* 2535. Drawn by J.C. Manning.

3. **Hesperantha ballii** Wild in Kirkia **4**: 136 (1963). —Goldblatt in Ann. Missouri Bot. Gard. **73**: 135 (1984). Type: Zimbabwe, Chimanimani Mts., Point 71, vii.1961, *Ball* 948 (SRGH, holotype).

Plants 12–25 cm high. Corms conical-campanulate, 9 mm in diameter at the widest; tunics light-brown, brittle-papery, irregularly lobed along the lower margin. Foliage leaves 4, the lower 2 basal and laminate, linear, reaching to about the base of the spike, the upper 2 leaves cauline and sheathing. Stem erect, unbranched. Spike 1–2-flowered; floral bracts green, usually flushed with red, 11–14 mm long, subequal or the inner shorter, the margins of the outer bract united around the stem for c. 3 mm. Flowers white, the outer tepals dark-red on the reverse, opening at sunset and then fragrant, downcurved, facing the ground; perianth tube 11–15 mm long, curving near the apex, usually slightly exceeding the bracts; tepals 15–19 × 5–6 mm. Anthers c. 7 mm long, yellow, pendent. Style branches c. 12 mm long. Capsules unknown.

Zimbabwe. E: Chimanimani Mts., NW slopes of Binga, 2200 m, 2.vii.1961, *Phipps* 3332 (SRGH). Restricted to the Chimanimani Mts., growing in thin soils on rock shelves and in grassland.
Related to *H. longicollis* Baker and the South African species *H. radiata* (Jacq.) Ker Gawl. and having a similarly long curved perianth tube, but easily distinguished from them by its short stature and 1–2-flowered spike.

16. SCHIZOSTYLIS Backh. & Harv.

Schizostylis Backh. & Harv. in Bot. Mag. **90**: t. 5422 (1864).

Perennial herbs with short rhizome-like rootstocks, aerial parts dying back annually. Leaves several, narrowly lanceolate, plane. Stem erect, usually unbranched. Inflorescence a distichous spike; floral bracts herbaceous, the inner smaller than the outer. Flowers actinomorphic, hypocrateriform; tepals united in a long cylindrical perianth tube, subequal. Stamens symmetrically arranged, the filaments exserted. Ovary globose, style dividing at the mouth of the perianth tube; style branches filiform, long and spreading, stigmatic along their entire length. Capsules membranous, oblong to globose. Seeds numerous per locule, more or less angular.

A genus of one species, *S. coccinea*, which extends from the eastern Cape Province of South Africa to eastern Zimbabwe, occurring in montane areas in wet habitats. It is closely related to *Hesperantha* and differs only in the rhizome-like rootstock. Plants sometimes produce cormlets in the upper leaf axils of a shape typical for *Hesperantha* corms, suggesting the rhizome is secondary in the genus.

Schizostylis coccinea Backh. & Harv. in Bot. Mag. **90**: t. 5422 (1864). —Baker in F.C. **6**: 56 (1896). —Letty, Wild Fl. Transvaal: 84, pl. 40, fig. 3 (1962). TAB. **18** fig. B. Type from South Africa (Cape Province).
Schizostylis pauciflora Klatt in Linnaea **35**: 380 (1867). —Baker in F.C. **6**: 56 (1896). Syntypes from South Africa (Natal, Orange Free State and Transvaal).

Plants 40–70 cm high. Rhizome erect, short. Leaves several, narrowly lanceolate (linear), mostly basal, reaching to about the base of the spike, 5–10 mm wide, the margins and midribs lightly thickened; upper leaves inserted on the stem well above the ground becoming progressively shorter, sometimes entirely sheathing. Stem usually unbranched, often producing corm-like propagules in the aerial axils. Spike (3)8–12-flowered; floral bracts green, 15–30 mm long, the inner about two-thirds as long as the outer. Flowers red, occasionally pink, rarely white; perianth tube (22)27–35 mm long, at least reaching the bract apices, more often exceeding them; tepals subequal, 25–35 × 10–15 mm, spreading horizontally and at right angles to the perianth tube. Filaments exserted 10–12 mm from the tube; anthers 8–10 mm long. Style branches 18–22 mm long, laxly spreading from the mouth of the perianth tube. Capsules oblong to obovoid, 10–12 mm long.

Zimbabwe. E: Nyanga, Gairezi R., xii.1958, *Davies* 2535 (K; MO; SRGH). **Mozambique**. MS: Tsetserra, 3.iii.1954, *Wild* 4496 (K).
Also in South Africa (Cape Province, Natal, Orange Free State and Transvaal), Lesotho and Swaziland. In wet places, especially mountain streams and seeps.
Grown as an ornamental especially in Europe, western N America and New Zealand.

17. ROMULEA Maratti

Romulea Maratti, Pl. Romul. Saturn.: 13, t. 1 (1772) nom. conserv. —de Vos in J. S. African Bot., Suppl. 9: 49 (1972); in Fl. Southern Africa **7**, 2(2): 10 (1983).

Perennial herbs with small globose corms, aerial parts dying back annually; corms with woody to firm membranous tunics. Leaves few to several, the lower 2–3 entirely sheathing (cataphylls), green, membranous or firm; foliage leaves basal, 1–several, more or less filiform with the margins and raised midribs often transversely winged, the blade thus terete with 4 longitudinal grooves, occasionally the leaves nearly plane with margins and midrib slightly thickened or only the margins thickened and winged. Stem short, aerial, or subterranean and becoming aerial in fruiting plants, simple or branching above or below the ground. Flowers solitary and terminal, sessile, subtended by 2 opposed bracts; floral bracts green, often the margins membranous and pale or ferrugineous, occasionally the inner bract entirely dry. Flowers actinomorphic, cupulate (hypocrateriform in a few South African species), variously coloured, often yellow in the centre; tepals united in a short to long perianth tube, subequal, usually spreading above. Stamens symmetrically disposed; filaments erect, more or less contiguous, sometimes united; anthers diverging or contiguous. Ovary globose, style dividing at or above the level of the anthers; style branches short, usually divided for half their length. Capsules oblong. Seeds many per locule, more or less globose or lightly angled, hard, glossy or matt.

A genus of c. 95 species, with two main centres of distribution - the SW Cape Province in the south and the Mediterranean countries in the north. Also represented in the mountains of eastern Africa, in Ethiopia and the Sudan, and in the Canary Islands.

Romulea camerooniana Baker in J. Bot. **14**: 236 (1876). —Wickens, Fl. Jebel Marra: 158 (1976). TAB. **19** fig. A. Type from Cameroon.

   Romulea fischeri Pax in Bot. Jahrb. Syst. **15**: 150 (1892). Type from Kenya.
   Romulea campanuloides Harms in Bot. Jahrb. Syst. **19**, Beibl. 47: 28 (1894). —Baker in F.T.A. **7**: 345 (1898). —de Vos in J. S. African Bot., Suppl. 9: 207 (1972); in Fl. S. Africa **7**,2(2): 52 (1983). Type from Tanzania.
   Romulea alpina Rendle in J. Linn. Soc., Bot. **30**: 401 (1895). Type from Tanzania.
   Romulea thodei Schltr. in J. Bot. **36**: 318 (1898). Type from South Africa (Orange Free State).
   Romulea thodei subsp. *gigantea* de Vos in J. S. African Bot. **21**: 106 (1955). Type from South Africa (Natal).
   Romulea campanuloides var. *gigantea* (de Vos) de Vos in J. S. African Bot. Suppl. **9**: 209 (1972).

Plants 7–12 cm high excluding the leaves. Corms 8–15 mm in diameter, globose, tapering to a sharp point below; tunics red-brown, woody or cartilaginous. Foliage leaves 1–4, straight or curved, 1–1.5 mm in diameter, filiform. Stems 1–4 per plant, more or less erect. Flowers solitary per stem; floral bracts green, striate, 12–18 mm long, the margins especially of the inner sometimes membranous and pale, the inner as long as or slightly shorter than the outer. Flowers lilac to violet, white to yellow in the centre, the tepals darkly veined; perianth tube 7–8 mm long; tepals subequal, 12–20 mm long, lanceolate, erect in the lower part and curving outwards above. Stamens free; filaments 4–5 mm long, erect, exserted c. 1 mm from the wide part of the perianth tube; anthers c. 5 mm long. Style dividing opposite the middle to upper half of the anthers, the branches c. 1 mm long. Capsules 7–10 mm long.

Zambia. N: Mafinga Hills, 1950 m, 12.iii.1961, *Robinson* 4468 (K). **Zimbabwe.** E: Nyanga Distr., foot of Mt. Inyangani, 15.ii.1931, *Norlindh & Weimarck* 5055 (BM; LD; S; SAM). **Malawi.** N: Nyika Plateau, Chowo Rock, 2200 m, 4.ii.1978, *Pawek* 13764 (K; MAL; MO). S: Zomba Plateau, 26.ii.1977, *Grosvenor & Renz* 968 (K; MO; SRGH).

Extending northwards to Ethiopia and Jebel Marra in the Sudan, and southward to Lesotho and the E Cape Province in South Africa. In montane grassland, amongst rocks and on rock outcrops and cliffs; flowering mostly in December to March.

A large-flowered and robust variant has been described as var. *gigantea*, but height and flower size are so variable in some populations that recognising infraspecific taxa is of doubtful value.

This species is closely related to *R. keniensis* Hedberg from northern East Africa. The latter species is distinguished by having the inner bracts almost entirely membranous and streaked with brown. In *R. cameroonensis* the bracts are herbaceous with membranous margins.

Tab. 19. A. —ROMULEA CAMEROONIANA, habit (×⅔), front view of flower (× 1), from *Goldblatt* 9101. B. —RADINOSIPHON LEPTOSTACHYA, whole plant (×⅔), flower (× 1), from *Goldblatt* 5933. Drawn by J.C. Manning.

18. RADINOSIPHON N.E. Br.

Radinosiphon N.E. Br. in Trans. Roy. Soc. S. Africa **20**: 262 (1932). —Carter in Fl. Pl. Africa **35**: pl. 1384 (1962).

Perennial herbs with corms, aerial parts dying back annually; corms with membranous to slightly fibrous tunics. Leaves several, the lower 2–3 membranous and entirely sheathing (cataphylls); the foliage leaves narrowly lanceolate, plane with slightly raised midrib and margins. Stem simple, or 1–3-branched. Inflorescence a secund spike; floral bracts green or somewhat membranous to dry apically, the inner usually shorter than the outer. Flowers zygomorphic, the tepals united in a long perianth tube, the upper tepal largest and arched over the stamens. Stamens unilateral and arcuate, the filaments exserted from the perianth tube; anthers parallel. Style dividing opposite or beyond the anthers, style branches filiform. Capsules globose, membranous. Seeds 2–4 per locule, more or less globose.

A genus probably consisting of a single polymorphic species, or possibly 2–3 cryptic species, frequent in mountain areas, extending from the Transvaal in South Africa and Swaziland to southern Tanzania.
The affinities of *Radinosiphon* are uncertain but the genus is probably most closely allied to *Gladiolus* and *Hesperantha*. Absence of winged seeds and apically broadened style branches exclude it from *Gladiolus* with which it shares a chromosome number (2n = 30), unique to subfamily *Ixioideae*. A resemblance to *Anomatheca* is misleading. *Radinosiphon* lacks the divided style branches and characteristic leaf anatomy of *Anomatheca*, which has a chromosome number of 2n = 22.

Radinosiphon leptostachya (Baker) N.E. Br. in Trans. Roy. Soc. S. Africa **20**: 263 (1932). —Carter in Fl. Pl. Africa **35**: pl. 1384 (1962). TAB. **19** fig. B. Type from South Africa (Transvaal).
  Lapeirousia leptostachya Baker, Handb. Irid.: 170 (1892). Type as above.
  Lapeirousia holostachya Baker in Bull. Misc. Inform., Kew **1894**: 391 (1894). Type: Zambia, Fwambo, south of Lake Tanganyika, ii.1893, *Carson 14* (K, holotype).
  Tritonia acroloba Harms in Bot. Jahrb. Syst. **30**: 278 (1902). Type from Tanzania.
  Acidanthera leptostachya (Baker) N.E. Br. in Bull. Misc. Inform., Kew **1921**: 297 (1921).
  Acidanthera holostachya (Baker) N.E. Br. in Bull. Misc. Inform., Kew **1921**: 297 (1921).
  Acidanthera lomatensis N.E. Br. in Bull. Misc. Inform., Kew **1921**: 298 (1921). Type from South Africa (Transvaal).
  Radinosiphon holostachya (Baker) N.E. Br. in Trans. Roy. Soc. S. Africa **20**: 263 (1932).
  Radinosiphon lomatensis (N.E. Br.) N.E. Br. in Trans. Roy. Soc. S. Africa **20**: 263 (1932).
  Radinosiphon cameronii N.E. Br. in Trans. Roy. Soc. S. Africa **20**: 263 (1932). Type: Malawi, summit of Chiradzulu, 1905, *Cameron 179* (K, holotype).

Plants (8)15–50 cm high. Corms 12–15 mm in diameter; tunics pale, usually membranous. Foliage leaves several, mostly basal, reaching to about the base of the spike, sometimes shortly exceeding the stems, 8–15 mm wide, narrowly lanceolate, plane with slightly raised midrib and margins, cauline leaves progressively reduced in size above. Stem simple or 1–3-branched. Spike (2)4–12-flowered or more; floral bracts 8–12 mm long, green or somewhat membranous to dry apically, the inner bracts about two-thirds as long as the outer. Flowers pink to purple; perianth tube 20–30 mm long, narrowly cylindric; tepals unequal, the upper 10–12 mm long, arched over the stamens, the lower 8–11 mm long, narrower than the upper, streaked with darker-pink in the midline. Filaments 5–6 mm long, exserted c. 3 mm from the perianth tube; anthers 3–5 mm long. Style dividing opposite or beyond the anthers, style branches filiform. Capsules 4–5 mm in diameter, globose.

Zambia. N: Mbala Distr., hill above Ndundu, c. 1680 m, 11.ii.1957, *Richards 8137* (BR; K).
Zimbabwe. E: Nyanga, Nyamziwa Falls, 12.i.1951, *Chase 3675* (BR; K; LISC; MO; SRGH).
Malawi. N: Viphya Plateau, c. 60 km SW of Mzuzu, 23.ii.1975, *Pawek 9102* (K; MAL; MO). C: Dedza, Ciwawo (Chiwao) Hill, 3.ii.1967, *Salubeni 537* (K; LISC; MAL). S: Blantyre Distr., Ndirande Mt., 3.iii.1970, *Brummitt 8855* (EA; K; MAL). **Mozambique**. N: Lichinga (Vila Cabral), near cruzeiro do serra do Massangulo, 1600 m, 25.ii.1964, *Torre & Paiva 10829* (LISC). T: Tete Distr., Macanga, Mt. Furancungo, c. 1519 m, 15.iii.1976, *Pereira, Sarmento & Marques 1680* (BR; WAG). MS: Chimanimani Mts., Moribane, 1200 m, 2.iii.1907, *Johnson 239* (K).

Also in South Africa (Transvaal), Swaziland and southern Tanzania. In rocky grassland, often among boulders and cliffs in mountainous areas; flowering in December to March.
Carter 1961, (ibid.) united all the species under *R. leptostachya*. This seems advisable as there appear to be no important differences between populations that merit species separation. Plants from different populations do, however, maintain some variation in overall robustness and perianth size in cultivation, indicating that there is a genetic basis to some of the variation.

19. GLADIOLUS L.

Gladiolus L., Sp. Pl.: 36 (1753). —Baker in F.T.A. **7**: 360 (1898). —G.J. Lewis et al. in J. S. African Bot., Suppl. 10: 6–316 (1972). —Goldblatt & de Vos in Bull. Mus. Natl. Hist. Nat. sér. 4, B, Adansonia **11**: 417–428 (1989).

Antholyza L., Sp. Pl.: 37 (1753).

Petamenes Salisb. ex J.W. Loudon, Ladies' Flower-Gard., Ornam. Bulb. Pl.: 42, pl. 8,3 (1841).

Sphaerospora Sweet ex J.W. Loudon, loc. cit.: 66, pl. 14,1 (1841).

Acidanthera Hochst. in Flora **27**: 25 (1844).

Homoglossum Salisb. ex Baker in J. Linn. Soc., Bot. **16**: 161 (1878).

Oenostachys Bullock in Bull. Misc. Inform., Kew **1930**: 465 (1930).

Perennial herbs with corms, aerial parts dying back annually; corms with coriaceous to fibrous and reticulate tunics. Leaves few to several, the lower (2)3 entirely sheathing and mostly below ground (cataphylls); foliage leaves usually synanthous (more or less fully developed at flowering), or hysteranthous (developing after flowering on the same shoot or borne on separate shoots), few to several, basal or cauline, blades well developed or reduced and largely to entirely sheathing, lanceolate to linear and plane or filiform and terete, the margins, midrib and sometimes also other veins thickened and hyaline (margins sometimes winged). Stem terete, simple or branched. Inflorescence a spike with flowers secund or, in a few species, distichous); floral bracts usually green, soft to firm, sometimes dry and brown at anthesis, relatively large, the inner usually smaller than the outer. Flowers zygomorphic (actinomorphic in a few species from South Africa, Zaire and Madagascar); tepals united in a well developed, sometimes very long perianth tube; subequal, or unequal with the uppermost broader and arching to hooded over the stamens, the lower 3 narrower, shorter or longer than the upper. Stamens unilateral and arcuate; filaments included or exserted from the perianth tube; anthers unilateral (symmetrical in actinomorphic flowers). Style exserted; style branches simple, expanded above and sometimes apically bilobed. Capsules large and slightly inflated. Seeds usually many, with a broad membranous wing about the circumference (wingless in a few species).

A genus of c. 215 species, centred in southern Africa and extending through tropical Africa and Madagascar to Europe and the Middle East. 37 species occur in the Flora Zambesiaca region of which 5 are endemic.

A collection of *Gladiolus pretoriensis* Kunth from "Botswana", *Pole Evans & Ehrens* 1914, is from near Swartruggens, W Transvaal (South Africa), apparently collected by Ehrens alone. The species probably does not grow in Botswana.

1. Perianth tube about twice as long as the tepals (or longer) and exceeding the bracts (at least 50 mm long); flowers white to cream-coloured, often with red to purple marks on the lower tepals   -   -   -   -   -   -   -   -   -   -   -   -   - 2
– Perianth tube shorter to slightly longer than both the upper tepal and the bracts (never twice as long); flowers variously coloured including white to cream   -   -   -   - 4
2. Blades of flowering-stem leaves short or lacking; leaf sheaths villous; bracts less 15–20 mm long   -   -   -   -   -   -   -   -   -   -   - 37. *curtifolius*
– Blades of flowering-stem leaves well developed, long; leaf sheaths glabrous; bracts 40–80 mm long   -   -   -   -   -   -   -   -   -   -   -   -   - 3
3. Perianth tube 50–90 mm long; lower tepals each with a red streak in the midline; anthers without acute apiculate apices   -   -   -   -   -   -   -   -   -   - 29. *bellus*
– Perianth tube (90)120–150 mm long; lower tepals with a narrow purple streak in the lower midline; anthers with apiculate apices 2–5 mm long   -   -   -   - 36. *callianthus*
4. Upper tepal 2–3 times as long as the other tepals; upper tepal twice as long as wide -   - 5
– Upper tepal shorter than, or up to 1.5 times as long as the other tepals; upper tepal rarely twice as long as wide   -   -   -   -   -   -   -   -   -   -   -   - 6
5. Perianth tube dimorphic, with a narrow cylindric lower part, abruptly expanded into a broadly cylindric upper part; upper tepal c. 15 mm long   -   -.   -   -   -   - 32. *huillensis*
– Perianth tube gradually expanding towards the throat, more or less infundibuliform; upper tepal 35–42 mm long   -   -   -   -   -   -   -   -   -   -   - 28. *magnificus*
6. Spike strongly distichous and stiffly erect; perianth usually pale with fairly uniform dark speckles   -   -   -   -   -   -   -   -   -   -   -   -   -   - 7
– Spike secund, erect or inclined; perianth rarely speckled, more often uniformly coloured or with contrasting marks on the lower tepals   -   -   -   -   -   -   -   - 8

7. Perianth greenish, covered with minute dark-red points; leaves linear, usually exceeding the spike, the margins and midribs (and sometimes other veins) heavily thickened and hyaline - - - - - - - - - 22. *sericeovillosus* subsp. *calvatus*
- Perianth whitish to cream-coloured, with minute grey-blue points; leaves lanceolate, usually shorter than the spike, the margins thickened and hyaline, the midribs and other veins more or less equal and not thickened - - - - - - - - - - - - 23. *elliotii*
8. Flowers generally large; perianth tube 15–45 mm long; upper tepal (18)24–55 mm long, 1–1.4 times as long as the perianth tube, (rarely shorter or more than twice as long as the perianth tube) - - - - - - - - - - - - - - - - - - 9
- Flowers generally small; perianth tube 3.5–14 mm long; upper tepal 8–24 mm long, rarely to 35 mm), 1.5–2 times as long as the perianth tube - - - - - - - - - 22
9. Floral bracts very large, 50–80 mm long, imbricate and 2–3 internodes long; upper tepal 2–2.5 times as long as the perianth tube - - - - - - - - - 24. *ecklonii*
- Floral bracts medium to large, 15–70 mm long, sometimes imbricate but not more than 2 internodes long - - - - - - - - - - - - - - - - 10
10. The lower 3 tepals about as long as, to slightly longer than the upper tepal (when the flower is viewed in profile); tepals extending outward; anthers with acute apiculate appendages c. 1.5 mm long - - - - - - - - - - - - - - - - - - - 11
- The lower 3 tepals two thirds to half as long as the upper tepal (when the flower is viewed in profile); tepals down-curved; anthers without acute apiculate appendages - - 13
11. Lower tepals without pale streaks; flowers medium in size, the upper tepal 18–20 mm long
35. *serenjensis*
- Lower lateral tepals each with a pale median streak; flowers large with the upper tepal 38–55 mm long - - - - - - - - - - - - - - - - - - 12
12. Perianth scarlet to dark-red with a spathulate yellow to cream streak on the lower lateral tepals; filaments exserted 20–25 mm from the perianth tube - - - - 33. *decoratus*
- Perianth pale to deep-pink with a more or less linear white to cream mark on the lower lateral tepals; filaments exserted 10(15) mm from the perianth tube - 34. *oligophlebius*
13. Perianth tube 25–40 mm long; upper tepal 35–50 mm long - - - - - 14
- Perianth tube 15–25 mm long; upper tepal 20–33 mm long - - - - - 16
14. Upper tepal arched over the stamens, not hooded and dipping below the horizontal; tepals white to cream-coloured, often finely veined with pink to red - - - - 21. *verdickii*
- Upper tepal hooded over the stamens, dipping below the horizontal; flowers in shades of red, orange, yellow, greenish or brown, without a fine network of darker veins - - 15
15. Leaves up to 35 mm wide and lightly to densely pubescent; leaf margins, midrib and 3 or more secondary veins heavily thickened and hyaline; upper tepal 30–40 mm long; tepals greenish-yellow - - - - - - - - - - - - - - - 26. *velutinus*
- Leaves (5)10–20(30) mm wide; margins and midrib lightly to moderately thickened and hyaline; upper tepal 25–35 mm long; tepals in various shades of red, orange, yellow or greenish
25. *dalenii*
16. Stamen filaments included in the perianth tube - - - - - - - 17
- Stamen filaments exserted 4–10 mm from the tube - - - - - - - 18
17. Plants normally with fully developed foliage leaves on the flowering stem; upper tepal 18–21 mm long - - - - - - - - - - - - - 30. *benguellensis*
- Plants normally with only sheathing leaves on the flowering stem, laminate foliage leaves produced on separate shoots after flowering; upper tepal 25–38 mm long - 31. *melleri*
18. Upper tepal hooded over the stamens, dipping below the horizontal; flowers in shades of cream to yellow - - - - - - - - - - - - - - - 27. *nyasicus*
- Upper tepal arched over the stamens, not hooded and dipping below the horizontal; flowers variously coloured - - - - - - - - - - - - - - - 19
19. Upper tepal 20–23 mm long; leaves well developed, 7–12 mm wide; (from southern Mozambique) - - - - - - - - - - - - 2. *hollandii*
- Upper tepal (24)25–35 mm long; leaf blades well developed or reduced, but never more than 6 mm wide; (not confined to southern Mozambique in the Flora Zambesiaca area) - 20
20. Tepals white to cream-coloured (pale-yellow), nearly always with pink to red or purple veins evident even when dry - - - - - - - - - - - 20. *erectiflorus*
- Tepals solidly pink, reddish or purple, sometimes with darker nectar guides on the lower tepals - - - - - - - - - - - - - - - - - - 21
21. Perianth tube 10–15 mm long; floral bracts 10–15 mm long; widespread elsewhere in the Flora Zambesiaca area but not in southern Mozambique or Botswana - - 10. *laxiflorus*
- Perianth tube 16–20 mm long; floral bracts 15–25 mm long; in the Flora Zambesiaca area only in southern Mozambique - - - - - - - - - - 11. *brachyphyllus*
22. Foliage leaves only two, the lower leaf sheathing the lower half of the stem, with a blade shorter than the sheath; the upper leaf shorter than the lower; stem flexed outwards at 45° above the upper leaf sheath - - - - - - - - - - - - - - - 23
- Leaves of the flowering stem (including those that are entirely sheathing) more than two; the lowermost not sheathing more than half the stem and the sheath exceeding the blade or not; stem more or less stiffly erect, or barely inclined above - - - - - - 24

23. Perianth tube 7–12 mm long; upper tepal at least 15(25) mm long   18. *gracillimus*
  – Perianth tube 3.5–5 mm long; upper tepal 8–12 mm long   -   -   -   -   19. *pusillus*
24. Leaves lanceolate, more than 5 mm wide, blades or at least the sheaths sparsely to heavily
    pubescent   -   -   -   -   -   -   -   -   -   -   -   -   -   -   -   25
  – Leaves either linear, less than 3 mm wide, or if wider and lanceolate then blades and sheaths
    glabrous   -   -   -   -   -   -   -   -   -   -   -   -   -   -   -   26
25. Leaves visibly (to the naked eye) fairly densely villous on the veins; perianth white with a
    dark-purple tube; upper tepal c. 20 mm long   -   -   -   -   -   5. *intonsus*
  – Leaves sparsely villous, sometimes only on the sheaths (indumentum sometimes visible only
    with a hand-lens); perianth pink to red; upper tepal 30–35 mm long   10. *laxiflorus*
26. Bracts (20)25–45 mm long, imbricate and 2(3) internodes long; stem unbranched; spike stiffly
    erect   -   -   -   -   -   -   -   -   -   -   -   -   -   -   -   27
  – Bracts either 10–16(20) mm long and not or barely imbricate, or 20–30(35) mm long, but not or
    hardly imbricate and less than 2 internodes long; stem branched or unbranched; spikes erect
    or inclined   -   -   -   -   -   -   -   -   -   -   -   -   -   -   28
27. Largest leaves lanceolate, (6)9–16(24) mm wide; seeds with well developed wings; anthers 7–10
    mm long   -   -   -   -   -   -   -   -   -   -   -   -   13. *gregarius*
  – Leaves linear or nearly so, 1.5–3(5–7) mm wide; seeds without wings; anthers 5–6 mm long
                                                                        14. *microspicatus*
28. Leaves more than 2 at the base of the flowering stem; blades well developed, reaching and
    sometimes exceeding the spike in length   -   -   -   -   -   -   -   -   29
  – Leaves usually evenly spaced on the flowering stem; blades short to absent, or sometimes leaves
    2 at stem base with blades reaching to the middle of the stem   -   -   -   -   35
29. Perianth whitish to cream with grey-blue markings and the apices drawn into linear appendages;
    leaves linear   -   -   -   -   -   -   -   -   -   9. *permeabilis* subsp. *edulis*
  – Perianth variously coloured, if whitish the apices not drawn into linear appendages; leaves
    lanceolate or linear   -   -   -   -   -   -   -   -   -   -   -   30
30. Leaves lanceolate, sometimes narrowly so, at least 4 mm wide   -   -   -   -   31
  – Leaves linear and usually less than 3 mm wide with thickened margins and midribs, the blade
    thus sometimes oval in section with narrow longitudinal grooves   -   -   -   33
31. Upper tepal 16–19 mm long; perianth pale yellowish-green   -   -   -   4. *flavoviridis*
  – Upper tepal 20–26 mm long, or more; perianth in shades of pink with darker nectar guides
                                                                        32
32. Leaf sheaths glabrous   -   -   -   -   -   -   ,   -   -   -   1. *crassifolius*
  – Leaf sheaths minutely pubescent   -   -   -   -   -   -   -   -   3. *zambesiacus*
33. Leaf sheaths minutely pubescent; perianth purple-pink   -   -   -   6. *zimbabweensis*
  – Leaf sheaths glabrous   -   -   -   -   -   -   -   -   -   -   -   -   34
34. Leaf blades with the midribs heavily thickened and thus oval-round in section; perianth purple
    to pink   -   -   -   -   -   -   -   -   -   -   -   -   -   -   7. *juncifolius*
  – Leaf blades with the midribs only slightly thickened; perianth red to orange with yellow nectar
    guides   -   -   -   -   -   -   -   -   -   -   -   -   -   -   8. *rubellus*
35. Tepals subequal and more or less symmetrically disposed; filaments included within the
    perianth tube   -   -   -   -   -   -   -   -   -   -   -   -   17. *muenzneri*
  – Tepals unequal, the upper largest and arching over the stamens, the lower 3 narrower than the
    upper; filaments partly exserted   -   -   -   -   -   -   -   -   -   -   36
36. Spike erect; corm tunics reddish-brown and membranous to finely fibrous; flower in profile
    windowed (gaping between the lower dorsal tepals and the remaining 5)   12. *unguiculatus*
  – Spike inclined; corm tunics straw-coloured and coarsely fibrous, the fibres mostly vertical and
    usually thickened below into claws; flowers not windowed in profile   -   -   -   37
37. Perianth white, usually flushed with purple and with dark-purple nectar guides, or uniformly
    dark-purple (violet); the tepals curving outwards in the upper half except the uppermost, and
    widely gaping   -   -   -   -   -   -   -   -   -   -   -   -   15. *atropurpureus*
  – Perianth yellow to yellow-brown, the lower tepals with yellow nectar guides; tepals directed
    forwards and not or hardly gaping   -   -   -   -   -   -   -   - 16. *serapiiflorus*

1. **Gladiolus crassifolius** Baker in J. Bot. **14**: 334 (1876); in F.C. **6**: 150 (1896). —Brenan in Mem. N.Y.
    Bot. Gard. **9**,1: 85 (1954). —G.J. Lewis et al. in J. S. African Bot., Suppl. 10: 60 (1972). TAB. **20** fig.
    A. Type from South Africa (Transvaal).
      Gladiolus thomsonii Baker, Handb. Irid.: 223 (1892); in F.T.A. **7**: 372 (1898). Type from
    Tanzania.
      Gladiolus masukuensis Baker in Bull. Misc. Inform., Kew **1897**: 283 (1897); in F.T.A. **7**: 365
    (1898). Type: Malawi, Misuku Hills (Mesuku Mts.) [type label], "Masuku Plateau," [protologue],
    (cultivated in Zomba), *Whyte* s.n. (K, lectotype here designated, the sheet with a branched plant
    and well-preserved flowers; K, isolectotype).
      Gladiolus mosambicensis Baker in F.T.A. **7**: 576 (1898). Type: Mozambique, Beira, without date,
    Braga 117 (B, holotype).
      Gladiolus gazensis Rendle in J. Linn. Soc., Bot. **40**: 210 (1911). —Eyles in Trans. Roy. Soc. S.
    Africa **5**: 331 (1916). —Plowes & Drummond, Wild Fl. Rhodesia: pl. 37 (1976). —Tredgold &
    Biegel, Rhod. Wild Fl.: 11, pl. 8 (1979). Type: Zimbabwe, Chimanimani (Melsetter), 1800 m,

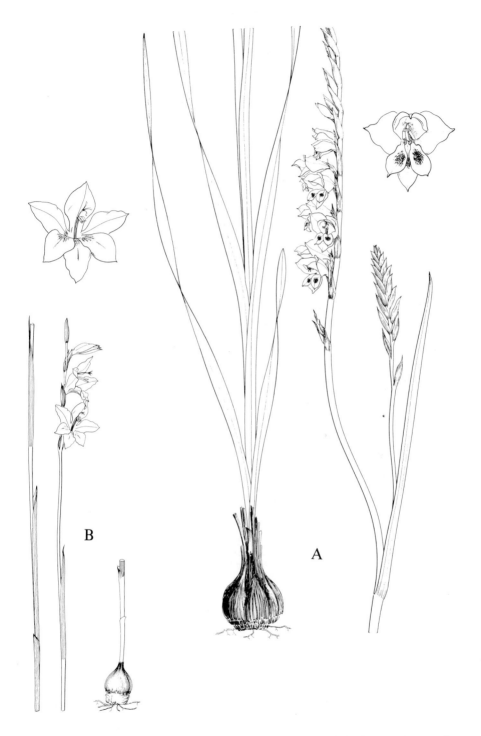

Tab. 20. A. —GLADIOLUS CRASSIFOLIUS, whole plant ($\times\frac{1}{2}$), flower ($\times\frac{3}{4}$), from *Chapman & Chapman* 8364 and *Plowes* 2118. B. —GLADIOLUS BRACHYPHYLLUS, whole plant ($\times\frac{1}{2}$), flower ($\times\frac{3}{4}$), from *Müller* 753 and *Bredenkamp* 1206. Drawn by J.C. Manning.

23.ix.1907, *Swynnerton* 779 (BM, lectotype here designated; K, isolectotype). Syntypes: Zimbabwe, Chirinda outskirts, 1200–1350 m, 1.ix.1907, and Mt. Pene, 2100–2200 m, 28.ix.1907; both labelled *Swynnerton* 779 and mounted on the same sheet as the lectotype (BM).
    Gladiolus unguiculatus sensu Tredgold & Biegel, Rhod. Wild Fl.: 12, pl. 8 (1979).

Plants (25)40–90(120) cm high, sometimes with up to 5 stems per corm. Corms 2–3 cm in diameter; tunics coriaceous to coarsely fibrous. Foliage leaves 4–8, either few and emergent at flowering time, then the stem bearing only 1–2 cauline and sheathing leaves and 2 laminate basal leaves, or with several basal laminate leaves and 2–3 cauline leaves, these with progressively shorter blades and the upper sometimes entirely sheathing; blades narrowly lanceolate to linear, exceeding the spike in short-stemmed plants, or reaching to about the base of the spike in tall plants, 5–12 mm wide, the midrib and margins hyaline and moderately thickened. Stem simple or 1–3(5)-branched. Spike with the main axis (6)12–22-flowered, the branches with fewer flowers; bracts 12–18 mm long, usually pale-green or flushed purple, soft-textured initially, becoming dry and membranous above (in fruit often translucent light red-brown), more or less obtuse to subacute, usually apiculate, the inner slightly shorter than the outer and 2-apiculate. Flowers pale to deep-pink or purple, the lower lateral and sometimes the lowermost tepals each with a dark band of colour across the lower half of the limbs, sometimes the dark band edged with cream; perianth tube c. 8 mm long, curved and funnel-shaped, cylindric below, widening and curving outward in upper part; tepals unequal, the uppermost 20–22 × c. 8 mm (often less when dry), larger than the others and arched over the stamens, the upper lateral tepals 18–22 mm long, directed forward and curving outward in the upper half, the lower 3 more or less straight and directed downward, united basally for 2–3 mm, the lateral tepals 12–15 mm, and the lowermost 16–20 mm long. Filaments c. 12 mm long, exserted for 6 mm; anthers 7–8 mm long. Style usually dividing opposite the lower half of the anthers; style branches c. 2 mm long, seldom extending past the middle of the anthers. Capsules 9–12(14) mm long, obovoid.

Zimbabwe. C: Makoni Distr., near Headlands, 15.x.1964, *Plowes* 2118 (K; SRGH). E: Mutare Distr., Vumba Mts., c. 1900 m, 16.ii.1964, *Chase* 6024 (BM; BR; K; LISC; S; SRGH). S: Bikita Distr., summit of Mt. Buhwa, 30.x.1973, *Biegel, Pope & Gosden* 4333 (C; MO; SRGH; WAG). **Malawi.** S: Mt. Mulanje, Sombani, halfway up Namasili, c. 2400 m, 16.v.1981, *Chapman & Patel* 5704 (BR; K; MAL). **Mozambique.** Z: Gurué, toward Namuli, 12.viii.1949, *Andrada* 1860 (COI; LISC). MS: Gorongoza Mts., near Morombodzi Falls, 10.viii.1944, *Mendonça* 2455 (BM; C; LISC; MO). M: Matutuíne, 15 km from Ponta de Ouro, 29.xii.1978, *de Koning* 7333 (LMU; WAG).
Especially common in eastern Zimbabwe and on Mt. Mulanje in Malawi; also in SW Tanzania, South Africa, Swaziland and Lesotho, and rare and localised in W Angola. Usually above 1800 m in the Flora Zambesiaca area, but at lower altitudes further south, occurring on the coastal plain in S Mozambique and South Africa. In open grassland, in well watered or seasonally wet sites; flowering in (August) September to November, especially after fires, or toward the end of the rainy season, March to June.

2. **Gladiolus hollandii** L. Bolus in J. Bot. **69**: 261 (1931). —G.J. Lewis et al., J. S. African Bot., Suppl. 10: 56 (1972). Type from South Africa (Transvaal).
    Gladiolus varius var. *elatus* F. Bolus in Ann. Bolus Herb. **2**: 104 (1917); I. Verd. in Fl. Pl. S. Africa **20**: t. 791 (1940). Type from South Africa (Transvaal).

Plants 75–120 cm high, usually growing in small clumps. Corms c. 30 mm in diameter; tunics of coarse brown fibres. Foliage leaves 6–8, the lower 4–5 basal, reaching to the base or middle of the spike, 12–22 mm wide, broadly to narrowly lanceolate, the margins slightly thickened, the midribs less so, upper leaves smaller than the basal and mostly sheathing. Stem erect, occasionally 1-branched. Spike 20–40-flowered; bracts green at anthesis, soon turning brown, 17–30 mm long, acuminate to setaceous. Flowers white with red to pink speckles and lines, the tube dark-red; perianth tube 18–25 mm long, obliquely infundibuliform, slender below, expanded apically; tepals subequal, 20–25 mm long, the uppermost slightly larger and ascending, the lower curving down and outward. Filaments 15–18 mm long, exserted c. 10 mm; anthers c. 7 mm long. Style dividing near the anther apices, style branches c. 2.5 mm long. Capsules 1–2 cm long, obovate.

Mozambique. M: Libombo Mts., 6–9 km from Goba on the way to Fonte dos Libombos, 31.iii.1945, *Estèves de Sousa* 122 (PRE).
Local in southern Mozambique, also in Swaziland and South Africa (Transvaal). In stony grassy slopes at altitudes of about 800 m; flowering in February to April.

3. **Gladiolus zambesiacus** Baker, Handb. Irid.: 212 (1892); in F.T.A. **7**: 364 (1898). —Gomes e Sousa, Subsid. Estudo Fl. Niassa Port.: 56 (1935) as "*zambeziacus*". —Brenan in Mem. N.Y. Bot. Gard. **9**: 84 (1954). Type: Malawi, Shire Highlands, near Blantyre, 1887, *Last* s.n. (K, lectotype, designated by D. Geerinck on the sheet, the more complete specimen).

Gladiolus buchananii Baker, Handb. Irid.: 212 (1892); in F.T.A. **7**: 362 (1898). Type: Malawi, summit of Ndirandi (Direndi), 1500 m, *Buchanan* s.n. (K, holotype; E).

Plants 45–75 cm high. Corms c. 15 mm in diameter; tunics of papery to finely fibrous layers. Foliage leaves 6–7, the lower 4 more or less basal and larger than the others, reaching to the base or middle of the spike, (2)3.5–6 mm wide, narrowly lanceolate to nearly linear, the sheaths especially of the lowermost leaves usually sparsely or sometimes densely short-pubescent, the blades occasionally so, the midrib and margins slightly thickened; upper leaves cauline and decreasing in size above, the uppermost usually entirely sheathing. Stem simple or 1(2)-branched. Spike 6–12-flowered, flexed below the first flower and inclined toward the ground; bracts green or flushed purple, becoming dry after flowering, (20)25–35(45) mm long, lanceolate, the inner slightly shorter than the outer. Flowers pink to mauve, fading to whitish in the throat, the lower 3 tepals each cream-coloured in the lower half; perianth tube c. 10 mm long, obliquely funnel-shaped, curving outwards near the apex; tepals unequal, the upper 3 larger than the rest, 22–26 × c. 12 mm (often less when dry), the uppermost inclined over the stamens, lanceolate, the lower 3 inclined toward the ground, the lower lateral tepals c. 18 mm long. Filaments 10–12 mm long, exserted 4–5 mm; anthers 7–8(10) mm long. Style arching behind the filaments, usually dividing at or beyond the anther apices; style branches c. 4 mm long, extending beyond the anthers, expanded above. Capsules 15–20 mm long, obovoid-ellipsoid.

Malawi. S: Zomba Plateau, near turn to Chingwe's Hole, 16.iii.1970, *Brummitt* 9167 (K; MAL). **Mozambique**. Z: Gurué Mt., road to Namuli Peak, c. 1300 m, 21.ii.1966, *Torre & Correia* 14721 (LISC).

Restricted to the higher elevations of the Shire Highlands of southern Malawi, and local in adjacent Mozambique. In exposed rocky montane habitats, often in thin soil in rocky crevices which is seasonally wet but well drained during the rainy season and bakes completely dry during the winter months; flowering mostly in March and April, sometimes in May.

This species is distinguished by the moderate-sized pink flowers with cream-coloured throats and cream-coloured lower tepals tipped with pink, by the yellow anthers, and by the several well developed leaves which are present at flowering time and which are nearly always minutely pubescent on the basal leaf sheaths and the lower blades.

The only collection seen so far from Mozambique has unusually narrow and entirely glabrous leaves, c. 3 mm wide with thickened margins and midribs. Its identity is uncertain. It is similar to *G. erectiflorus* but lacks the dark-pink to red-veined tepals which distinguish that species.

4. **Gladiolus flavoviridis** Goldblatt sp. nov. Plantae 55–85 cm altae, foliis 6, inferioribus duobus basalibus longissimis laminibus sublinearibus marginibus costisque leviter incrassatis, foliis superioribus brevioribus, spicis 10–18-floris, floribus flavis vel flavo-viridibus, tepalis inaequalibus, superioribus grandioribus 15–17 mm longis, filamentis c. 10 mm longis 6 mm exsertis, antheris 5.5 mm longis. Typus: Zimbabwe, Mutare, north of Christmas Pass (Buru's grounds), 15.ii.1949, *Chase* 1364 (SRGH, holotype; BM; K).

Plants 55–85 cm high. Corms unknown. Foliage leaves 6, the lower 2 usually more or less basal and larger than the others, reaching to the base or middle of the spike, 4–5 mm wide, nearly linear, the midrib and margins slightly thickened; upper leaves cauline and progressively shorter above, the uppermost usually entirely sheathing. Stem simple or with a short branch. Spike 10–18-flowered, inclined from the base; bracts evidently membranous below and dry and brown above at anthesis, 12–16 mm long, the inner about two-thirds as long as the outer. Flowers pale-yellow or greenish, tepal markings if present unknown; perianth tube c. 10 mm long, obliquely infundibuliform, widening and curving outwards near the apex, c. 10 mm long; tepals unequal, the upper three 15–17 mm long, (often less when dry), larger than the others, broadly lanceolate, the uppermost inclined over the stamens, c. 12 mm wide, the lower 3 tepals united below for c. 2 mm, 9–10 mm long, abruptly expanded in the upper half, the apices inclined toward the ground. Filaments c. 10 mm long, exserted 6 mm; anthers c. 5.5 mm long. Style arching over the filaments, dividing between the base and middle of the anthers; style branches 2–3 mm long, never reaching the anther apices. Capsules c. 10 mm long, obovoid.

Zimbabwe. E: Mutare, ridge south of the Customs Border Station, 20.ii.1958, *Chase* 6821 (K; SRGH).

Apparently restricted to the Mutare area of eastern Zimbabwe, and probably also in adjacent Mozambique. In deciduous woodland or open grassland; flowering in January to early March.

This species is distinguished by the relatively small, pale-yellow or greenish flowers on a ± deflexed secund spike and bracts which by anthesis are dry and rust-coloured above.

5. **Gladiolus intonsus** Goldblatt sp. nov. Plantae 50–75 cm altae, foliis 6–8 inferioribus 2–3 basalibus longissimis superioribus caulinis, laminis lanceolatis-sublinearibus 7–12 mm latis pubescentibus, marginibus costatisque leviter incrassatis hyalinisque, spicis 9–14-floris, bracteis 16–20(25) mm longis viridibus supra siccis, floribus albis, cremeis vel leviter purpureis flavinotatis, tepalis inaequalibus superioribus c. 20 mm longis grandissimus, inferioribus unguiculatis angustis, filamentis c. 12 mm longis c. 7 mm exsertis, antheris c. 9 mm longis. Typus: Malawi, Nyika Plateau, 15 km north of M1, 1830 m, 11.iii.1978, *Pawek* 14066 (MAL, holotype; BR; K; MO; PRE; SRGH).

Plants 50–75 cm high. Corms c. 15 mm in diameter, globose; tunics brown coriaceous, the outer layers breaking into medium to coarse vertical fibres. Foliage leaves 6–8, the lower 2–3 basal and larger than the others, reaching to just above the base of the spike, 7–12 mm at the widest, narrowly lanceolate to sublinear, usually densely (sometimes lightly) pubescent, the margins and midrib slightly thickened; the upper leaves cauline and progressively reduced in size. Stem unbranched, glabrous. Spike 9–14-flowered; bracts evidently green below, becoming dry and light-brown above, 16–20(25) cm long, narrowly lanceolate, attenuate; the inner about two-thirds as long as the outer. Flowers white to cream-coloured (sometimes flushed with pink) or light-purple, the lower lateral tepal or all 3 lower tepals each with a broad yellow median streak, the lower part of the tube dark-purple; perianth tube c. 10 mm long, obliquely infundibuliform; tepals unequal, the uppermost larger than the others, c. 20 × c. 12 mm, obovate, the lower 3 tepals spreading more or less horizontally, c. 20 × c. 8 mm, lanceolate, narrowed into claws below. Filaments c. 12 mm long, exserted for c. 7 mm; anthers c. 9 mm long. Style dividing opposite the apex of the anthers; style branches c. 2.5 mm long. Capsules unknown.

Zambia. N: east of Mbala, Mbosi Road, c. 1650 m, 31.iii.1932, *Thompson* 1103A (K). **Malawi**. N: 8 km NE of Mpora, c. 1800 m, 29.iii.1976, *Phillips* 1604 (K; MO); Chitipa Distr., 15 km E of crossroads to Karonga, Songea stream, 19.iv.1969, *Pawek* 2253 (MAL; K).

Also in Tanzania and SE Zaire. In woodland in hilly country; flowering in late summer and autumn, March to early May.

Recognised by its densely pubescent leaves and cataphylls, and small white to cream-coloured (rarely pink to light-purple) flowers.

6. **Gladiolus zimbabweensis** Goldblatt sp. nov. Plantae 20–35 cm altae, foliis 4–5, omnibus laminatis, inferioribus grandioribus, laminis subteretibus 4-sulcatis, vaginis saepe pubescentibus, spica (4)8–12-floris, floribus caeruleis vel roseis vel malvinis atrorosei- vel atropurpurei-notatis, tepalis inaequalibus, superioribus grandioribus, filamentis 8–9 mm exsertis, antheris c. 6 mm longis. TAB. 21 fig. A. Typus: Zimbabwe, Nyanga Distr., slopes of Mt. Inyangani, c. 2100 m, 16.ii.1964, *Chase* 8123 (SRGH, holotypus; K; UPS; WAG).

Plants 20–35 cm high. Corms 9–12 mm in diameter; tunics of fine pale netted fibres, sometimes extending upward in a neck. Foliage leaves 4–5, the lower 2 basal, reaching to about the base of the spike, 1–1.5 mm wide, linear, the margins and midrib strongly thickened and raised, thus arching over the surface, edges of the raised parts ciliate to pubescent; the upper leaves cauline and progressively shorter, the uppermost sheathing for half of its length; sheaths, especially of the lower leaves, lightly pubescent, distinctly ribbed when dry. Stem erect, rarely branched, flexed at or slightly above the middle. Spike 4–8(12)-flowered, inclined toward the ground; bracts green to purplish especially above, 10–15 mm long, the inner somewhat shorter than the outer. Flowers pale grey-blue or purple to pink, the lower lateral tepals each with a pale transverse band outlined in purple across the lower part of the limb; perianth tube 7–10 mm long, curving outward between the bracts, widening gradually toward the mouth; tepals unequal, the uppermost 14–19 × c. 10 mm, larger than the others, arched to hooded over the stamens, the upper lateral tepals smaller, directed forward and curving outward near the apices, the lower 3 tepals c. 10 × c. 4 mm, straight, inclined, joined to the upper laterals for 2–4 mm, usually slightly exceeding the uppermost when viewed in profile, fairly abruptly expanded in the upper half, the lower lateral tepals channelled below. Filaments 12–14 mm long, exserted for 8–9 mm; anthers c. 6 mm long, with an apiculum 0.3–0.5 mm long. Style dividing opposite the lower third of the anthers; style branches c. 2 mm long, entwined with the anthers. Capsules 12–15 mm long, obovoid.

Zimbabwe. E: Mt. Engwa, c. 1980 m, 8.ii.1955, *Exell, Mendonça & Wild* 294 (BM; LISC; SRGH). **Mozambique**. MS: Tsetserra, 2140 m, 7.ii.1955, *Exell, Mendonça & Wild* 246 (BM; LISC; SRGH).

Tab. 21. A. —GLADIOLUS ZIMBABWEENSIS, whole plant (×¾), flower, dissected and front views (×1), from *Goldblatt* 9080. B. —GLADIOLUS JUNCIFOLIUS, whole plant (×¾), flower (×1) from *Grosvenor* 177 and *Taylor* 1790. Drawn by J.C. Manning.

Restricted to the highlands of eastern Zimbabwe and neighbouring Mozambique. In rocky, well watered, montane grassland; flowering in mid January to April.

This species may be distinguished by its very small flowers, together with its linear leaves heavily thickened on the margins and midribs, and the light pubescence on the leaf sheaths.

7. **Gladiolus juncifolius** Goldblatt sp. nov. Plantae 20–35 cm altae, foliis 4–5, inferioribus laminatis superioribus vaginantibus, laminis teretibus 4-sulcatis, spica (2)3–6-flora, floribus roseis vel malvinis flavinotatis, tepalis inaequalibus, superioribus grandioribus, filamentis 5 mm exsertis, antheris c. 6.5 mm longis. TAB. **21** fig. B. Typus: Zimbabwe, Chimanimani Mts., 2200 m, 8.x.1960, *Plowes* 2129 (SRGH, holotypus; K).

Plants 20–35 cm high. Corms unknown. Foliage leaves 4–5, the lower 2–3 basal and with long blades reaching at least to the base of the spike, sometimes exceeding it, (basal leaves occasionally burned off and then apparently lacking); sheaths, especially the lower, conspicuously ribbed; blades 1–1.5 mm wide, terete, the margins and midrib heavily thickened (the blades thus narrowly longitudinally 4-grooved), edges of the raised parts sometimes ciliate; upper leaves 2–3, inserted above ground and entirely sheathing, decreasing in size above, widely spaced, the uppermost 15–30 mm long. Stem erect, unbranched, usually straight. Spike (2)3–6-flowered; bracts completely green or dry and brownish above, 9–14 mm long, the inner somewhat shorter than the outer. Flowers pale-pink or purplish, the 2 lower lateral tepals each with a yellow transverse band across the lower part of the limb and darker-pink distally; perianth tube c. 9 mm long, curving outward between the bracts, widening gradually toward the mouth; tepals unequal, the uppermost c. 20 mm long, larger than the others and arched over the stamens, the upper lateral tepals smaller, directed forward and ultimately curving outwards, the 3 lower tepals horizontal or down-curved, joined for c. 3 mm, limbs fairly abruptly expanded in the upper half, the lowermost tepal c. 18 × c. 7 mm, the lower lateral tepals c. 15 × c. 3 mm. Filaments c. 15 mm long, exserted by 5 mm; anthers c. 6.5 mm long with an apiculum c. 0.5 mm long. Style dividing toward the apex of the anthers; style branches c. 3 mm long. Capsules unknown.

Zimbabwe. E: Chimanimani Mts., near Mountain Hut, 24.ix.1966, *Grosvenor* 177 (K; LISC; SRGH; UPS).

Restricted to the highlands of eastern Zimbabwe and probably also in neighbouring Mozambique, mostly recorded from the Chimanimani Mountains but also from Mt. Inyangani to the north. In rocky montane grassland, evidently in sites that usually retain moisture throughout the year; flowering in mid July to October.

This species may be distinguished by its relatively small pale-pink flowers, erect spike and terete 4-grooved foliage leaves, the latter being produced long before the flowers and are sometimes dry or broken at flowering time. The foliage leaves are quite different from the short, entirely sheathing cauline leaves borne on the flowering stem, a distinction unusual in the genus. The species is unusual in that it flowers in the dry season.

8. **Gladiolus rubellus** Goldblatt sp. nov. Plantae 25–45 cm altae, foliis 3–4, inferioribus 1–2 basalibus linearibus (1.8)3–4.5 mm latis, marginibus costisque incrassatis hyalinis, foliis superioribus brevioribus, caule saepe ramoso, spica 10–16-flora, bracteis 12–14(16) mm longis, floribus rubro-aurantiacis tepalis inferioribus flavi-striatis, tubo perianthii c. 14 mm longo, tepalis inaequalibus superioribus 16–20 mm longis, inferioribus brevioribus, filamentis 8–10 mm exsertis, antheris c. 6.5 mm longis, seminibus alatis. Typus: Botswana, Molepolole, 20.v.1984, *Plowes* 7085 (UCBG, holotypus; MO; PRE).

Plants 25–45 cm high. Corms 15–20 mm in diameter, obconic; tunics of fairly fine vertical fibres. Foliage leaves 3–4, at least the lower 1–2 inserted at or below ground level, these longer than the others and usually slightly longer than the stem, occasionally shorter, (1.8)3–4.5 mm wide, linear, the midrib and margins hyaline and strongly thickened, margins raised at right angles to the surface; the upper 1–2 leaves cauline, shorter than the basal leaves, the uppermost sometimes entirely sheathing. Stem often 1(2)-branched. Spike 10–16-flowered, secondary spike with fewer flowers; bracts green, 12–14(16) mm long, the inner usually shorter than the outer. Flowers bright orange-red, the lower lateral tepals each with a transverse yellow band outlined in dark-red across the distal half; perianth tube c. 14 mm long, widening and curving outwards near the apex; tepals unequal, lanceolate, the uppermost 16–20 mm long (shorter when dry), larger than the others, the 3 lower tepals 12–14 mm long, joined with the upper lateral tepals for 2–4 mm, and to themselves for c. 3 mm, exceeding the uppermost tepal when viewed in profile, abruptly expanded in the upper half, more or less horizontal, flexed downwards distally. Filaments 12–14 mm long, exserted for 8–10 mm; anthers c. 6.5 mm long. Style dividing between the middle and apex of the anthers, style branches 2.5–3 mm long. Capsules 8–10 mm long (possibly longer), obovoid.

Botswana. SE: c. 36 km WNW of Lobatse on Kanye road, 18.i.1960, *Leach & Noel* 192 (K; PRE). Restricted to south-eastern Botswana between Molepolole and Lobatse. In stony ground, often in *Acacia-Combretum* woodland; flowering in January to March, rarely as late as May.

9. **Gladiolus permeabilis** D. Delaroche, Descr. Pl. Aliq. Nov. **27**: t. 2 (1766). —Eyles in Trans. Roy. Soc. S. Africa **5**: 331 (1916). —Suessenguth & Merxmüller, Contrib. Fl. Marandellas Distr.: 76 (1951). Type a plant cultivated in Leiden.

Subsp. **edulis** (Burch. ex Ker Gawl.) Oberm. in J. S. African Bot., Suppl. 10: 135 (1972). Type from South Africa (Cape Province).
    Gladiolus edulis Burch. ex Ker Gawl. in Bot. Reg. **2**: t. 169 (1817); Baker in F.C. **6**: 161 (1896); in F.T.A. **7**: 373 (1898). —Sölch in Merxm. Prodr. Fl. SW. Afrika, fam. 155: 4 (1969). Type as above.
    Gladiolus remotiflorus Baker in Bull. Herb. Boissier, sér. 2, **1**: 867 (1901). Type from South Africa (Transvaal).

Plants 30–60 cm high. Corms 10–12 mm in diameter; tunics coarsely fibrous. Foliage leaves 3–6, at least the lower 2–3 inserted at or below ground level, these shorter than or about as long as the stem, occasionally longer, 1–2.5(3) mm wide, linear, the midrib and margins slightly thickened and hyaline; the upper 2–3 leaves cauline, shorter than the basal leaves and sometimes entirely sheathing. Stem simple or branched. Spike 4–8-flowered; bracts apparently green to grey below, membranous and becoming dry and pale above, 7–12 mm long, the inner usually shorter than the outer, the outer acute to somewhat attenuate, the inner forked at the apex. Flowers usually intensely sweet-scented, whitish to dull-cream, grey or mauve, flushed with grey-blue on the upper tepal, the keels dark-purple to maroon, the lower lateral tepals yellow in the upper half; perianth tube 8–12 mm long, widening and curving outward near the apex; tepals unequal, all long-attenuate and narrowed into claws below, the uppermost 16–20 mm long, larger than the others, arched over the stamens, the lower 3 joined for c. 2 mm with the upper lateral tepals, and to themselves for 3–5 mm, 13–15 mm long, the limbs abruptly expanded, more or less horizontal, or flexed downward distally, the cusps often twisted. Filaments 13–14 mm long, exserted for 9–10 mm; anthers 5–6 mm long. Style dividing between the middle and apex of the anthers; style branches 1.5–2 mm long, much expanded above. Capsules 6–10 mm long, nearly globose to obovoid.

Botswana. N: near Tshesebe (Tsessebe) Station, 6.i.1974, *Ngoni* 247 (SRGH). SE: SE of Ramotswa, 16.iv.1977, *Hansen* 3126 (C; PRE). **Zimbabwe.** W: Nyamandhlovu, Pasture Research Station, ii.1954, *Plowes* 1680 (K; SRGH). C: Rusape Distr., 21.ii.1940, *Hopkins* in GHS 7533 (SRGH). E: between Chipinge (Chipinga) and Mt. Selinda, c. 900 m, 16.ii.1962, *Chase* 7624 (SRGH). S: Masvingo (Ft. Victoria), 1909, *Monro* 841 (BM; SRGH).
    Also in South Africa (mostly N Cape Province and W Transvaal) and Namibia. Mostly in semi-arid parts of the Flora Zambesiaca region, usually on sandy soils, in grassland and *Acacia* woodland; flowering in February to April.
    This subspecies is distinctive in its linear leaves, small intensely fragrant pale flowers and long-apiculate tepals with dark median streaks. The upper tepal is narrow and strongly arched. When viewed in profile there is a gap or window between the uppermost tepal and the remaining tepals.
    Included here are plants from the Great Dyke in the Makonde Distr. of Zimbabwe (*Philcox & Müller* 9079 (K; SRGH)) which have leaves with heavily thickened margins, pink flowers and tepals which lack the characteristic tapering apiculae. Additional material is needed to establish whether this represents a separate taxon.

10. **Gladiolus laxiflorus** Baker in Trans. Linn. Soc. London, ser. 2, Bot. **1**: 268 (1878); Handb. Irid.: 211 (1892); in F.T.A. **7**: 362 (1898). —Wild in Fl. Pl. Africa **33**: t. 1287 (1959). —Geerinck in Bull. Jard. Bot. Belg. **42**: 279 (1972). Type from Angola.
    Gladiolus tritonioides Baker in Bull. Misc. Inform., Kew **1895**: 74 (1895); in F.T.A. **7**: 363 (1898). —Fries, Wiss. Ergebn. Schwed. Rhod.-Kongo-Exped.: 235 (1916). Type: Zambia, Fwambo, ix.1893, *Carson* 37 (K, holotype).
    Gladiolus antunesii Baker in F.T.A. **7**: 576 (1898). Type from Angola.
    Gladiolus trichophyllus Diels in Notizbl. Bot. Gart. Berlin-Dahlem **13**: 264 (1936). Type from Tanzania.

Plants (20)40–70 cm high. Corms 25–30 mm in diameter, depressed-globose, usually persisting for some years, usually dark-red; tunics coriaceous to somewhat membranous, fragmenting irregularly. Foliage leaves 3–6, the lower 2–4 basal and laminate, the blades beginning to emerge around flowering time, ultimately the largest up to 35 cm long, 4–6 mm wide, linear to narrowly lanceolate or falcate, the margins and midribs moderately thickened, usually lightly pubescent, especially on the sheaths; the upper 1–3 leaves mostly to entirely sheathing, glabrous. Stems often 1–2-branched, occasionally simple. Spike 6–9-flowered, the lateral spikes with fewer flowers, inclined; bracts 10–15(20) mm long, green below, becoming more or less membranous above or dry at anthesis, the inner usually slightly longer than the outer. Flowers purple to pinkish, without markings, or 1 or all of the lower tepals with a small yellow mark in the distal half; perianth tube c. 10 mm long, expanding gradually from base to apex, curving outward above; tepals subequal to unequal with the upper 3 larger than the lower ones, 30–35 × 13–15 mm at the widest (poorly pressed flowers may shrink by up to 50 per cent of their original size), lanceolate, narrowed below, the upper 3 ascending, the lower directed downward. Filaments 18–20 mm long; exserted for 12–14 mm; anthers 7–8 mm long. Style dividing beyond the anther apices, style branches c. 3 mm long. Capsules 15–22 mm long, ellipsoid-obovoid.

Zambia. B: Kaoma (Mankoya), c. 82 km west of Kafue Hoek pontoon, 7.xi.1959, *Drummond & Cookson* 6212 (E; K; LISC; MO; PRE; SRGH). N: c. 16 km from Kawambwa on Nchelenge road, 29.xi.1961, *Richards* 15447 (BR; K; SRGH). W: Kitwe, 17.xi.1967, *Mutimushi* 2362 (K; NDO). C: Kabwe (Broken Hill), 3.ix.1938, *Pole Evans & Erens* 1906 (K; P). **Malawi**. N: Nyika Plateau, Chelinda Bridge, c. 2300 m, 24.xi.1986, *la Croix* 855 (K; MAL; MO).

Also in Angola, Zaire (Shaba) and SW Tanzania. In wet sites such as marshes, dambos and stream banks; flowering in October and November, sometimes into late December, just before or with the first rains when the surrounding vegetation is fairly short.

This species may be distinguished by its fairly large deep-pink to reddish-purple flowers and by its short leaves usually covered with a light pubescence, at least below. Poorly preserved specimens with the flowers distorted and shrunk to half their size may be confused with *G. unguiculatus* Baker. However, the latter has glabrous, sheathing leaves on the flowering stem, long bladed foliage leaves produced on separate shoots, unbranched stems, and smaller flowers with unequal tepals nearly always with very dark spade-shaped markings on the lower lateral tepals.

11. **Gladiolus brachyphyllus** F. Bolus in Ann. Bolus Herb. **2**: 103 (1917). —G.J. Lewis et al. in J. S. African Bot., Suppl. 10: 291 (1972). TAB. **20** fig. B. Type from South Africa (Transvaal).

Plants 60–100 cm high. Corms 15–20 mm in diameter; tunics membranous or becoming irregularly fibrous, reddish-brown. Foliage leaves of the flowering stem 3–4, long-sheathing, the blades to 4–8 cm long and shorter than the sheaths, lanceolate, the margins and midribs hyaline and usually slightly thickened, not imbricate; hysteranthous foliage leaves on separate shoots are not known and are probably not produced; leaves of non-flowering plants 2–3, up to 80 × 0.5–1.5 cm, linear, margins usually slightly thickened. Stem erect, simple or 1-branched. Spike 3–10-flowered, flexed at the first flower and inclined; bracts green, soft-textured, 15–25(30) mm long, lanceolate to attenuate, the inner slightly shorter than the outer. Flowers pink to purple, marked with dark-reddish spots at the base of the lower tepals; perianth tube 16–20 mm long, obliquely funnel-shaped, emerging between the bracts; tepals unequal, the uppermost arcuate, 24–30 mm long, 12–14 mm wide, the upper lateral tepals smaller, directed forwards and ultimately curving outwards, the 3 lower tepals horizontal to down-curved, 18–22 × 7–9 mm at the widest, lanceolate, not narrowed below into claws, about as long as the uppermost when viewed in profile. Filaments (16)18–22 mm long, exserted 10–12 mm; anthers 8–10 mm long. Style dividing near the anther apices, style branches 3–4 mm long. Capsules 20–25 mm long, obovoid.

Mozambique. M: Boane, road to Goba, 60 m, 16.x.1980, *Schäfer* 7291 (K; LMU; MO; NBG).

Also in South Africa (E Transvaal) and probably Swaziland. In lightly wooded grassland, (or low montane grassland in South Africa), sometimes in rocky soils and rock outcrops; flowering in November–December, at the end of the dry season or early in the wet season.

This species is characterised by its comparatively short leaves, sometimes almost entirely sheathing, and a bright red-purple flower with dark-red to purple markings at the bases of the lower tepals.

12. **Gladiolus unguiculatus** Baker in J. Linn. Soc., Bot. **16**: 178 (1877) as nom. nov. pro. *G. cochleatus* Baker nom. illegit. non Sweet (1832); Handb. Irid.: 233 (1892); in F.T.A. **7**: 372 (1898). —Mildbraed in Bot. Jahrb. Syst. **58**: 232 (1923). —Hepper in F.W.T.A. ed. 2, **3**: 144 (1968). —G.J. Lewis et al. in J. S. African Bot., Suppl. 10: 292 (1972). —Wickens, Fl. Jebel Marra: 158 (1976). —Blomberg-Ermatinger & Turton, Some Fl. Pl. SE. Botswana: pl. 124 (1988). TAB. **22** fig. A.
Gladiolus cochleatus Baker in J. Bot. **14**: 333 (1876) nom. illegit. non Sweet (1832). Type from Sierra Leone.

Gladiolus brevicaulis Baker in Trans. Linn. Soc. London, ser. 2, Bot. 1: 267 (1878). —Eyles in Trans. Roy. Soc. S. Africa 5: 331 (1916). Type from Angola.

Gladiolus oatesii Rolfe in Oates, Matabeleland, ed. 2: 410 (1889). —Baker, Handb. Irid.: 226 (1892); in F.T.A. 7: 373 (1898). —Fries, Wiss. Ergebn. Schwed. Rhod.-Kongo-Exped.: 236 (1916). —Eyles in Trans. Roy. Soc. S. Africa 5: 331 (1916). —Martineau, Rhod. Wild Fl.: 20, plate 13 (1954). Type: Zimbabwe, Matabeleland, without precise locality or date, *Oates* s.n. (K, holotype).

Antholyza labiata Pax in Bot. Jahrb. Syst. 15: 156, t. 7, figs. 1–4 (1892). —Baker in F.T.A. 7: 374 (1898). Type from Togo.

Antholyza cabrae De Wild. in Ann. Mus. Congo Belge sér. V, 1: 15 (1903). Type from Zaire.

Antholyza thonneri De Wild., Etudes Fl. Bangala et Ubangi: 208 (1911). Type from Zaire.

Gladiolus brevifolius sensu Eyles in Trans. Roy. Soc. S. Africa 5: 331 (1916) non Jacquin (1791).

Gladiolus thonneri (De Wild.) Vaupel in Mildbraed, Wiss. Ergebn. Deutsch. Zentr.-Afr.-Exped. 1910–1911, Bot.: 10, 67 (1922).

Gladiolus labiatus (Pax) N.E. Br. in Trans. Roy. Soc. S. Africa 20: 267 (1932).

Gladiolus cabrae (De Wild.) N.E. Br. in Trans. Roy. Soc. S. Africa 20: 267 (1932).

Gladiolus atropurpureus sensu Geerinck, Bull. Jard. Bot. Belg. 42: 271 (1972), in part; sensu Obermeyer, Fl. Pl. Africa 44: pl. 1760 (1977), in part, non Baker (1876).

Plants 30–60 cm high. Corms 15–25(35) mm in diameter, often dark-red on the outside and sometimes internally; tunics red-brown, membranous and irregularly broken or fibrous and reticulate. Leaves of the flowering stem are usually short, entirely stem sheathing and hardly distinguishable from the cataphylls, or these sometimes with a short isobilateral blade, rarely do 2 basal leaves become fairly well developed and with blades exceeding the sheaths; sheathing leaves 3–5, usually 6–9 cm long, the lower ones exceeding the internodes, the upper ones shorter than the internodes, thus imbricate below but rarely so above, sometimes the lower ones with blades up to 4 cm long; (1)2–3 foliage leaves are produced on separate shoots after the plants have flowered (hysteranthous leaves), these ultimately 30–45 cm long and 4–8(12) mm wide, linear to narrowly lanceolate, with margins and midrib slightly thickened and hyaline. Stem erect, rarely branched. Spike 10–18-flowered; bracts green, 10–15 mm long, the inner somewhat shorter than, to nearly equalling, the outer in length. Flowers cream to light-purple, the upper tepals flushed light- to deep-purple, the lower each with deep-purple spear-shaped marking in the upper third, surrounded by a lighter area, sometimes with a dark spot at the base of the uppermost tepal, flowers windowed when viewed in profile, with a gap between the uppermost tepal and the upper lateral tepals; perianth tube c. 10 mm long, curving outward from between the bracts, widening near the mouth; tepals unequal, the uppermost 18–20(24) × (8)10–12 mm, larger than the others and arched over the stamens and style and much narrower toward the base, the upper lateral tepals smaller, joined to the lower 3 for 3–5 mm, directed forward and ultimately curving outwards, the lower 3 tepals smaller than the others, usually exceeding the uppermost when viewed in profile, 10–12 mm long, usually united for 1–2 mm, horizontal or directed downward distally, narrowed below into claws, the limbs abruptly expanded. Filaments 10–12 mm long, exserted 4–5 mm; anthers yellow, 6–8 mm long. Style dividing opposite the lower half of the anthers; style branches 2–2.5 mm long, reaching to between the middle and apices of the anthers. Capsules 12–16 mm long, ellipsoid-ovoid.

Botswana. SE: rocky hill c. 10 km N of Lobatse, c. 1350 m, 23.x.1977, *O.J. Hansen* 3242 (K; PRE; SRGH; UCBG; UPS; WAG). **Zambia**. B: c. 80 km E of Makoya on road to Kafue Hoek, 21.xi.1947, *Drummond & Cookson* 6718 (K; LISC; PRE). N: Kasama Distr., c. 6 km from Kalolo Village on path to Chambeshi Flats, 9.xii.1964, *Richards* 19324 (B; BR; K). W: Solwezi, 12.x.1953, *Fanshawe* 387 (BR; K; NDO). S: vlei near Kalomo, 16.xii.1961, *Whellan* 1894 (K; LISC; SRGH). **Zimbabwe**. N: Hurungwe, xi.1956, *Davies* 2251 (BR; K; LISC; PRE; SRGH). W: Bulawayo, ii.1902, *Eyles* 1224 (BM; SRGH). **Malawi**. N: Katoto, c. 5 km W of Mzuzu, 7.xi.1970, *Pawek* 3943 (K; MAL). C: by the M1 just N of Kasungu, 1100 m, 27.xii.1986, *la Croix* 4247 (K; MAL; MO). S: Zomba Distr., Mbidi Estate, 11.i.1960, *Banda* 365 (BM; MAL; SRGH). **Mozambique**. N: between Namatil and Nampula, 2.xii.1936, *Torre* 1087 (COI; LISC). Z: Pebane, 43 km from Pebane to Mualama, 15.i.1968, *Torre & Correia* 17153 (LISC).

One of the most widespread *Gladiolus* species, extending from Senegal and Gambia to Sudan and southwards to Angola and South Africa (Transvaal). Fairly common throughout its distribution range and a conspicuous element of the prerain or early wet season flora. In seasonally wet sites with shallow soils and poor drainage, often found in seepage areas on rock outcrops and along the margins of dambos; flowering at the end of the dry season or early in the wet season, in October to December in south tropical Africa.

The absence of foliage leaves on the flowering stem combined with an erect spike of small flowers distinctively windowed in profile makes this species easy to recognise. The flowering stem bears (2)3 fairly short, non-overlapping sheathing leaves, while true foliage leaves are produced on separate shoots from the same corm toward the end of the flowering cycle.

Tab. 22. A. —GLADIOLUS UNGUICULATUS, whole plant (× $\frac{1}{2}$), flower, lateral view (× 1), from *Schaijes* 5123. B. —GLADIOLUS GREGARIUS, whole plant (× $\frac{1}{2}$), flower, dissected and front views (× 1), from *Richards* 19040 and *Billiet & Jadin* 4204. Drawn by J.C. Manning.

This species is often confused with *G. atropurpureus* which has flowers of a similar size and colouring and also lacks foliage leaves on the flowering stem. However, spikes of *G. atropurpureus* are always inclined and the flowers have broader tepals and are not windowed in profile. The corms of *G. atropurpureus* also differ in their smaller size and their coarsely netted tunics which are often thickened below into claw-like ridges and therefore unlike the rather large corms of *G. unguiculatus* with their reddish, membranous to finely fibrous tunics.

Plants from the southern part of the species distribution range (Botswana, W Zimbabwe and the Transvaal) appear to differ significantly from typical *G. unguiculatus*. Corresponding to *G. oatesii* Rolfe, they have smaller corms and the lower leaves consistently have short blades and, as far as it is possible to tell from herbarium material, do not produce foliage leaves from separate shoots. An argument can be made for recognising these populations as a separate species or subspecies but they may equally be no more than a local race. There is not enough information available at present to make an informed decision.

13. **Gladiolus gregarius** Welw. ex Baker in Trans. Linn. Soc. London, ser. 2, Bot. **1**: 268 (1878); Handb. Irid.: 210 (1892); in F.T.A. **7**: 365 (1898). —Hepper in Kew Bull. **21**: 493 (1968); in F.W.T.A. ed. 2, **3**(1): 144 (1968). —Geerinck in Bull. Jard. Bot. Belg. **42**: 276 (1972). TAB. **22** fig. B. Type from Angola.

    Gladiolus spicatus Klatt in Linnaea **35**: 377 (1867) nom. illegit. non *G. spicatus* L. (1753). Type from Nigeria.

    Gladiolus multiflorus Welw. ex Baker in Trans. Linn. Soc. London, ser. 2, Bot. **1**: 269 (1878); Handb. Irid.: 221 (1892); in F.T.A. **7**: 369 (1898). Type from Angola.

    Gladiolus hanningtonii Baker, Handb. Irid.: 212 (1892); in F.T.A. **7**: 366 (1898). Type from Tanzania.

    Gladiolus karendensis Baker in Bull. Herb. Boissier, sér. 2, **1**: 867 (1901). Types from Tanzania.

    Gladiolus uhehensis Harms in Bot. Jahrb. Syst. **28**: 365 (1901). Type from Tanzania.

    Antholyza gilletii De Wild. in Ann. Mus. Congo Belge, sér. IV, **1**: 19 (1902). Type from Zaire.

    Antholyza descampsii De Wild. in Ann. Mus. Congo Belge, sér. IV, **1**: 18 (1902). Types from Zaire (Shaba).

    Gladiolus corbisieri De Wild. in Fedde, Repert. Spec. Nov. Regni Veg. **12**: 296 (1913). Types from Zaire (Shaba).

    Gladiolus elegans Vaupel in Bot. Jahrb. Syst. **48**: 536 (1913). Type from Tanzania.

    Gladiolus klattianus Hutch. in F.W.T.A. ed. 1, **2**: 379 (1936) as nom. nov. pro *G. spicatus* Klatt.

    Gladiolus pseudogregarius Mildbr. ex Hutch. in F.W.T.A. ed. 1, **2**: 379 (1936) nom. illegit. sine descr. lat.

Plants (15)30–80 cm high. Corms (12)18–30(40) mm in diameter; tunics membranous, fragmenting irregularly, occasionally subfibrous and the fibres then somewhat clawed. Foliage leaves 4–7, the lower 3–5 more or less basal, half to two thirds as long as the stem and (6)9–16(24) mm wide, narrowly lanceolate, the margins and midribs not or hardly thickened; upper leaves cauline and shorter. Stem erect and straight, unbranched. Spike (2)8–20(25)-flowered, erect, often congested, evidently straight but flexuous under the bracts; bracts 20–30(50) × 8–10 mm at the widest, green or becoming dry and brown at the end of flowering, firm, sometimes slightly striate (the veins raised and hyaline), imbricate, 2–2.5(3) internodes long, the inner c. two-thirds as long as the outer. Flowers usually light-to dark-purple (dark-reddish brown) fading to cream-coloured in the throat or mostly white, the tube usually dark-purple below, the lower tepals each with a dark-purple diamond-shaped mark in the distal third; perianth tube c. 12 mm long, curving outward and widening above; tepals unequal, the uppermost (16)22–26 × 12–18 mm, arched over the stamens, the lower 3 tepals c. 12 × c. 5 mm, usually shortly exceeding the uppermost when viewed in profile, narrowed below into claws, the limbs flexed downward and channelled, joined to the upper lateral tepals for 4–6 mm and to one another for c. 2 mm. Filaments 9–13 mm long, exserted 5–6 mm; anthers 7–10 mm long. Style dividing opposite the lower half of the anthers, style branches c. 2 mm long. Capsules 12–15 mm long, narrowly ellipsoid, fairly hard and woody.

Zambia. B: Mongu, 20.i.1966, *Robinson* 6806 (B; K). N: edge of Kalambo Falls Gorge, c. 1200 m, 8.ii.1965, *Richards* 19603 (K; MO; P). W: Ndola, 27.ii.1954, *Fanshawe* 867 (BR; K; NDO; SRGH). C: Mkushi Distr., c. 42 km from Kapiri Mposhi to Serenje, 3.ii.1973, *Strid* 2893 (C; K; MO). C: Luangwa Valley near Kapamba River, 3.iii.1970, *Astle* 5793 (K; NDO; SRGH). **Malawi**. N: Karonga Distr., Vinthukutu Forest Reserve, 3 km north of Chilumba, 26.iv.1975, *Pawek* 9565 (K; MAL; MO). C: Dzalanyama Forest Reserve near Chionjeza (Chiungiza), 9.ii.1959, *Robson* 1516 (BM; BR; K; SRGH). **Mozambique**. N: Massangulo, 24.iii.1933, *Gomes e Sousa* 1351 (COI). T: Angónia, near Domué, 22.ii.1980, *Macuácua & Mateus* 1129 (WAG). Z: Gurué, 26 km from Mutuáli to Lioma, 10.ii.1964, *Torre & Paiva* 10506 (LISC).

Widespread in tropical Africa, from southern Zambia to Senegal. In open woodland, often in rocky habitats where it is sheltered from competition, occasionally recorded from wetter sites such as dambo margins. Flowering in January to March in south tropical Africa.

This species is striking in its long overlapping floral bracts that clasp the spike. The bracts are at least two internodes in length and normally 20–25 mm long. Plants are particularly variable in size and individuals within a single population can range in height from 25–70 cm, with a corresponding variation in the number of flowers in the spike.

14. **Gladiolus microspicatus** Duvign. & Van Bockstael ex Còrdova in Bull. Jard. Bot. Belg. **60**: 326 (1990). Type from Zaire.
    Gladiolus klattianus subsp. *angustifolius* Van Bockstael in Bull. Soc. Roy. Bot. Belgique **96**: 126 (1963). Type from Zaire.

Plants 24–35 cm high. Corms 10–20 mm in diameter; tunics of fine to coarse yellow-brown to grey fibres, usually accumulating in a dense mass, often extending upward and with the dry cataphylls forming a thick neck. Foliage leaves (3)4–5, the lower (2)3–4 more or less basal, reaching to at least the base of the spike or sometimes the lower leaves shortly exceeding it, 1.5–3(7) mm wide, linear (rarely lanceolate), the margins not thickened; the upper leaves decreasing in size above, the uppermost usually cauline. Stem erect, unbranched. Spike 4–10-flowered, inflexed at the base, straight, sheathed by the bracts; bracts 18–25 mm long, green, or becoming dry and brown apically, firm and erect, clasping the stem, usually imbricate, 1–2 internodes long, the inner c. two-thirds as long as the outer. Flowers light- to dark-purple fading to cream-coloured in the throat and tube, the lower tepals each with a yellow median streak in the upper third; perianth tube 10–12 mm long, curving outward and widening above; tepals unequal, the uppermost c. 18 × 10–12 mm, arched over the stamens, the lower 3 tepals usually shortly exceeding or equalling the uppermost in length when viewed in profile and united with the upper lateral tepals for c. 4.5 mm and to each other for c. 2 mm, narrowed below into claws, horizontal or the limbs flexed downward distally. Filaments 10–12 mm long, exserted 5–6 mm from the perianth tube; anthers 5–6 mm long, usually purple, pollen yellow. Style dividing opposite the lower third of the anthers, style branches 1–1.5 mm long. Capsules 10–12 mm long, narrowly elliptic, comparatively hard and woody.

Zambia. N: Mbereshi-Kawambwa road, c. 1050 m, 18.i.1960, *Richards* 12404 (BR; K).
Centred in southern Zaire and there most common in rocky grassland in heavy-metal enriched soils, also in W Tanzania and Burundi. In well drained grassland or rock outcrops often in poor sandy soils; flowering in January to February.
This species resembles a diminutive *G. gregarius*, but differs in having shorter bracts, fewer flowers per spike, firm narrow leaves rarely more than 3–4 mm wide, a neck of coarse fibres around the base and, at least where known, wingless seeds. Some plants from Burundi have uniformly dark-purple flowers and bracts sometimes less than 1 internode long.

15. **Gladiolus atropurpureus** Baker in J. Bot. **14**: 335 (1876); Handb. Irid.: 211 (1892); in F.T.A. **7**: 364 (1898). —Geerinck in Bull. Jard. Bot. Belg. **42**: 271 (1972) excl. *G. unguiculatus*. TAB. **23** fig. A. Type: Mozambique, Morrumbala (Zambezi highlands), 18.i.1863, *Kirk* s.n. (K, lectotype, designated by Geerinck, 1972).
    Gladiolus caerulescens Baker in Trans. Linn. Soc. London, ser. 2, Bot. **1**: 267 (1878); Handb. Irid.: 211 (1892); in F.T.A. **7**: 364 (1898). Type from Angola.
    Gladiolus luridus Welw. ex Baker in Trans. Linn. Soc. London, ser. 2, Bot. **1**: 267 (1878); Handb. Irid.: 211 (1892); in F.T.A. **7**: 365 (1898). Type from Angola.
    Gladiolus flexuosus Baker in Bull. Misc. Inform., Kew **1894**: 390 (1894); in F.T.A. **7**: 372 (1898) nom. illegit. non *Gladiolus flexuosus* L.f. (1782). Type: Zambia, Fwambo (Lake Tanganyika), i.1893, *Carson* 79 (K, holotype).
    Gladiolus whytei Baker in Bull. Misc. Inform., Kew **1897**: 282 (1897); in F.T.A. **7**: 363 (1898). Type: Malawi, Mt. Malosa, Nov.–Dec. 1896 *Whyte* s.n. (K, lectotype here designated, the specimen most complete and best matching the description; B, isolectotype).
    Gladiolus gracilicaulis G.J. Lewis in J. S. African Bot. **7**: 29 (1941). Type as for *G. flexuosus* Baker.

Plants 30–60 cm high. Corms (10)15–20(30) mm in diameter; tunics fibrous, pale straw-coloured, fibres mostly vertical and often thickened and claw-like below. Foliage leaves of the flowering stem 3–4(5), entirely sheathing or with blades to 5(10) cm long, imbricate, narrowly lanceolate to linear, the margins and midribs hyaline and usually slightly thickened (hysteranthous foliage leaves on separate shoots are not produced); leaves of non-flowering plants solitary, 15–20 cm long, 4–6(12) mm wide, linear, usually with slightly thickened margins. Stem erect, unbranched (rarely branched), usually slightly flexed above the sheath of the upper leaf. Spike (3)5–10(15)-flowered, slightly flexuous, inclined; bracts green, 10–15(20) mm long, attenuate, rather soft-textured, the margins often membranous, the inner bracts somewhat shorter to nearly as long as the

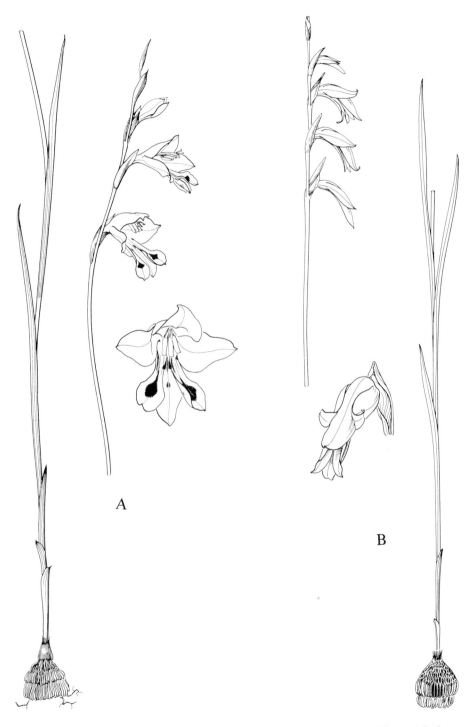

Tab. 23. A. —GLADIOLUS ATROPURPUREUS, whole plant (×$\frac{1}{2}$), front view of flower (×$\frac{3}{4}$), from *Goldblatt et al.* 8247. B. —GLADIOLUS SERAPIIFLORUS, whole plant (×$\frac{1}{2}$), individual flower (× 1), from *Schaijes* 2709. Drawn by J.C. Manning.

outer. Flowers cream-coloured to light-purple with the upper tepals flushed light- to deep-purple, the lower lateral tepals each with a deep-purple spade-shaped mark in the distal third surrounded by a lighter area or partly to entirely dark-purple, sometimes with pale markings on the lower tepals; perianth tube 10–12 mm long, curving outward between the bracts, widening near the mouth; tepals unequal, the uppermost 15–20 × 9–12 mm, larger than the others and arching over the stamens, the upper lateral tepals smaller directed forwards and curving outwards distally, the 3 lower tepals (10)12–15 × 6–9 mm and horizontal to down-curved, joined with the upper laterals for c. 5 mm and usually exceeding the uppermost when viewed in profile, narrowed below into claws, the limbs abruptly expanded. Filaments 6–8 mm long, exserted c. 4 mm; anthers 5–8 mm long. Style dividing opposite the upper half of the anthers, style branches 2–2.5 mm long, usually reaching nearly to the apices of the anthers, rarely exceeding them. Capsules (10)12–18 mm long, ovoid to ellipsoid.

Zambia. N: c. 24 km S of Chitipa and c. 6.5 km into Zambia, Nyanyunyi Village facing the Mafinga Hills, 1050 m, 27.xii.1972, *Pawek* 6161 (K). N: Mbala Distr., Ndundu, 1600 m, 1.xii.1960, *Richards* 13638 (K). W: Kitwe, 27.xii.1955, *Fanshawe* 2680 (K; LISC; NDO; S). C: Lusaka Distr., Chilanga, 1.i.1958, *Benson* 226B (BR; K). S: Choma Distr., dambo at Siamambo, 1.xii.1962, *Lawton* 1019 (K; NDO). **Zimbabwe**. N: Makonde Distr., Mhangura (Mangula), 3.xii.1961, *Jacobsen* 1536 (PRE; SRGH). C: Harare, 20.xii.1926, *Eyles* 4581 (K; SRGH). E: Mutare (Umtali) Golf Course, 26.xi.1957, *Chase* 6776 (COI; K; LISC; PRE; SRGH). **Malawi**. N: Livingstonia Escarpment, c. 980 m, 31.xii.1973, *Pawek* 7685 (K; MAL; MO). C: Dedza Distr., Chongoni, Ciwawo (Ciwao) Hill, 19.i.1959, *Robson & Jackson* 1259 (K; SRGH). S: Zomba Plateau, 27.xi.1976, *Goldblatt* 4512 (MO; WAG). **Mozambique**. N: Massangulo, i.1933, *Gomes e Sousa* 1233 (COI). Z: Morrumbala Mountain, west slopes, 9.xii.1971, *Pope & Müller* 569 (K; MO; PRE; SRGH). MS: Mavita, Serra Mocuta, 13.xii.1965, *Pereira & Marques* 1057 (LMU; WAG).

Also in Angola, Zaire (Shaba), Tanzania and Burundi. In light, deciduous woodland or low montane grassland, sometimes on rocky outcrops; flowering at the beginning of the wet season, November–December, sometimes even before soaking rains have fallen, and fruiting in late December and January and then becoming dormant through the wet months of February to April.

G. atropurpureus is often confused with *G. unguiculatus*, but may be distinguished from that species in its corm and flower morphology (see discussion under the latter). It is variable in the degree of development of the cauline leaves which may be entirely sheathing or have blades reaching the base of the spikes. Flowers range from deep maroon-purple to white with dark-purple nectar guides.

16. **Gladiolus serapiiflorus** Goldblatt sp. nov. Plantae 30–60 cm longae, foliis 4–6 imbricatis laminis reductis linearibus marginibus costisque incrassatis hyalinisque, spicis 4–8(12)-floris, bracteis 10–14 mm longis, floribus fusco-viridis, flavis vel brunneis, tepalis 10–13 mm longis, filamentis c. 8 mm longis 3–4 mm exsertis, antheris 6 mm longis. TAB. **23** fig. B. Typus: Zambia, Machili, woodland, 17.xii.1960, *Fanshawe* 5981 (K, holotypus; NDO; SRGH).

Plants 30–60 cm high. Corms 15–20 mm in diameter; tunics pale straw-coloured, fibrous with fibres reticulate or mostly vertical and then often thickened and claw-like below. Foliage leaves of the flowering stem 4–6, almost entirely sheathing, or the lower leaves laminate with blades 3–8 cm long, occasionally to 30 cm long and then reaching the top of the spike, linear (lanceolate), imbricate, margins and midribs hyaline and slightly thickened, upper leaves always predominantly sheathing; hysteranthus foliage leaves on separate shoots are not known and are probably not produced; leaves of non-flowering plants not known. Stem erect, unbranched, usually slightly flexed above the sheath of the upper leaf. Spike 4–8(12)-flowered; bracts green becoming dry above, 10–14 mm long, the inner somewhat shorter than to nearly as long as the outer. Flowers dull- to bright-yellow, the uppermost tepal and the limbs of the 2 lower lateral tepals sometimes brown, or the lower lateral tepals deeper yellow; perianth tube 7–8 mm long, curving outward between the bracts, widening near the mouth; tepals unequal, the uppermost c. 13 × 7–10 mm, larger than the others and arching over the stamens, the upper lateral tepals smaller c. 12 mm long and directed forward and curving upward apically, the 3 lower tepals c. 10 mm long, usually exceeding the uppermost when viewed in profile, united for 3 mm, down-curved, narrowed below into claws, the limbs channelled c. 3 mm at the widest. Filaments c. 8 mm long, exserted 3–4 mm; anthers c. 6 mm long. Style dividing near the anther apices, style branches c. 2.5 mm long. Capsules 12–15 mm long, ellipsoid.

Zambia. B: Machili, 17.xii.1960, *Fanshawe* 5981 (K; NDO; SRGH). W: Luanshya, 19.xii.1954, *Fanshawe* 1734 (NDO; K). C: c. 60 km E of Lusaka, Chipata (Fort Jameson) Road, 2.i.1958, *Benson* 225 (K).

Also in Zaire (Shaba). In light woodland or open low grassland, sometimes in rocky soils or on rock outcrops; flowering in November–December, at the end of the dry season or early in the wet season.

Resembling the common south tropical African species *G. atropurpureus* Baker in general appearance, mainly in the coarsely fibrous corm tunics and the weakly developed leaf blades on the flowering stems. *G. serapiiflorus* may be distinguished by its yellow to brownish flowers with long narrow tepals extended forward to form a tube-like structure with a narrow opening at the tepal apices. The flowers of *G. atropurpureus* are either purple or white with purple markings and the tepals are held fairly widely apart.

17. **Gladiolus muenzneri** Vaupel in Bot. Jahrb. Syst. **48**: 542 (1913). TAB. **24** fig. A. Type from Tanzania.

Plants 30–60 cm high. Corms (10)20–35 mm in diameter; tunics brownish, more or less coriaceous or fragmenting into coarse- to medium-textured fibres, sometimes entirely fibrous. Foliage leaves 2–3, entirely sheathing or with short occasionally imbricate blades, the blades 1–8(12) cm long, lanceolate usually narrowly so, margins and midribs hyaline and usually slightly thickened; (hysteranthus foliage leaves on separate shoots are not known and are probably not produced); leaves of non-flowering plants are not known. Stem erect, unbranched, usually straight. Spike 2–8-flowered; bracts green, or sometimes membranous or dry above, 18–25 mm long, the inner somewhat shorter than the outer and apically bifid. Flowers white to cream or yellow, sometimes flushed pinkish or light-mauve especially on the upper 3 tepals, or rarely uniformly mauve, pink or light-orange, lacking contrasting markings on the lower tepals; perianth tube 7–8 mm long, curving outward between the bracts, widening near the mouth; tepals nearly equal, 20–25 × (5.5)7–8 mm, directed forwards, weakly curving outward distally, hardly or not at all narrowed below into claws. Filaments 4–5 mm long, usually included in the perianth tube, exserted for up to 1 mm; anthers 7–9 mm long, sometimes the bases included for up to 2 mm, minutely apiculate. Style dividing toward the anther apices, style branches 3.5–4 mm long. Capsules 10–12 mm long, globose.

Zambia. N: Kawambwa Distr., Luapula Leper Settlement, by Mbereshi R., 2.xii.1961, *Richards* 15486 (K; SRGH). **Malawi**. N: Mzimba Distr., Champhila (Champira), c. 18 km SE of Katete over Lwanjati Pass, c. 1500 m, 25.xii.1975, *Pawek* 10515 (K; MAL; MO).

Also in SW Tanzania. In relatively dry deciduous woodland or in submontane grassland, usually above 1200 m; flowering in December until mid-January.

This species is recognised by its cream to pale-pinkish flowers without nectar guides, sometimes flushed with light-purple or apricot, subequal elongate tepals, and the short filaments which are almost to entirely included within the wide upper half of short perianth tube. The leaf blades are reduced to only a few centimetres in length or are vestigial, so that the leaves consist only of a sheath. Unlike *G. unguiculatus* Baker plants of this species do not produce foliage leaves from separate shoots later in the season.

18. **Gladiolus gracillimus** Baker in Bull. Misc. Inform., Kew **1895**: 74 (1895); in F.T.A. **7**: 363 (1898). —Geerinck in Bull. Jard. Bot. Belg. **42**: 275 (1972) pro parte. TAB. **24** fig. B. Type: Zambia, Fwambo (Lake Tanganyika), in 1893, *Carson* 118 (K, holotype).

Gladiolus aphanophyllus Baker in F.T.A. **7**: 363 (1898). Type from Tanzania, Tanganyika Plateau, [Urungu], *Carson* s.n. (K, holotype [mounted together with two spikes of *G. unguiculatus*]).

Plants small, 15–30(40) cm high. Corms 9–12 mm in diameter; tunics of fine netted fibres, occasionally extending upward in a weakly developed neck. Foliage leaves 2, the lower one basal and tightly sheathing the stem for half to three-quarters of its length with the blade usually reaching to about the top of the spike, the upper leaf inserted shortly below the spike and shorter or sometimes exceeding the apex of the lower leaf; blades 3–5 cm long and 0.5–1 mm wide, linear with 2 fairly narrow grooves on each surface, the margins and midrib somewhat to moderately thickened. Stem erect, unbranched, sheathed for most of its length, flexed sharply outward above the sheath of the upper leaf. Spike 2–4(7)-flowered, inclined at an angle of c. 45° or more, flexuous; bracts green, sometimes flushed purple, 7–10 mm long, the inner slightly shorter than the outer. Flowers usually pale-blue to lilac, occasionally white, the lower 3 tepals each with a dark-blue to purple diamond-shaped marking in the upper half, or flowers uniformly purple, or pale-yellow to greenish with deep-yellow markings on the lower 3 tepals; perianth tube 7–10 mm long, curved outward and emerging from between the bracts; tepals unequal, the uppermost (8)10–15 × 8–11 mm, larger than the others and arched over the stamens, the upper lateral tepals directed forward, the lower 3 tepals 7–12 ×

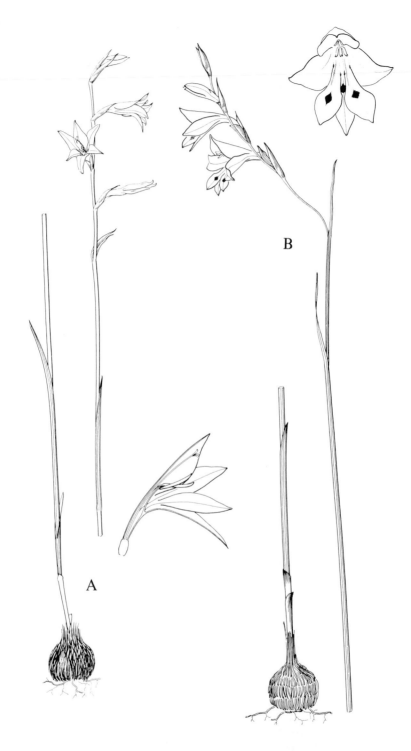

Tab. 24. A.—GLADIOLUS MUENZNERI, whole plant (×$\frac{1}{2}$), dissected flower (× 1), from *la Croix* 746. B.—GLADIOLUS GRACILLIMUS, whole plant (×$\frac{1}{2}$), front view of flower (×1), from *Goldblatt et al.* 8138. Drawn by J.C. Manning.

4–5 mm, much exceeding the upper when viewed in profile, united for 3–4 mm, held close together, more or less horizontal or tilted downward, abruptly narrowed below into claws. Filaments 6–8 mm long, usually exserted 3–4 mm; anthers 4.5–6 mm long. Style dividing at the apex of the anthers, style branches 1.5–2 mm long. Capsules 7–8 mm long, obovate-ellipsoid.

Zambia: N: Mbala Distr., near Nakatali Farm, 14.xi.1965, *Richards* 20655 (BR; K; LISC).
Malawi. N: between Chinunka and Ifumbo, c. 16 km E of Chitipa, 27.xii.1970, *Pawek* 4182 (K; MAL; SRGH).

Also in SW Tanzania. In seasonally waterlogged dambos, swampy places, or seasonally wet shallow soil over laterite; flowering in November to December, and occasionally into early January, at the end of the dry season or early in the wet season.

This species is distinctive in its pale-blue flowers borne on a strongly inclined spike. It is unusual in having only two foliage leaves, the lower of which sheaths the stem for most of its length and has a short linear blade that usually exceeds the short upper leaf. This vegetative form is also seen in the Zairean species *G. tshombeanus* Duvigneaud & Van Bockstael, and in the largely Zambian species *G. pusillus*. The latter may be distinguished from *G. gracillimus* by its very small, usually yellow flowers.

19. **Gladiolus pusillus** Goldblatt sp. nov. Plantae 20–30 cm altae, graciles, tunicis cormi fibrosis griseis, foliis 2 limbis brevibus linearibus, bracteis 6–9 mm longis, spica (2)4–9-flora inclinata, floribus flavis vel albis luteinotatis, tubo perianthii 3.5–4.5 mm longis, tepala superiore 8–10 mm longis, antheris 4.5–5 longis, seminibus alatis. Typus: Zambia, Kawambwa Distr., Ntumbachushi (M'tunatusha) R., 1290 m, 28.xi.1961, *Richards* 15422 (K, holotypus; BR; SRGH).

Plants slender, 15–30(40) cm high. Corms 7–12 mm in diameter; tunics of fine netted fibres sometimes accumulating around the base in a weakly developed neck. Foliage leaves 2, the lower one basal and tightly sheathing the stem for most of its length, often reaching to about the top of the spike, the upper leaf inserted shortly below the spike, the apex usually shorter than but occasionally exceeding the lower leaf; blades 2–5 cm long and c. 0.5 mm wide, linear and with 2 fairly narrow grooves on each surface, the margins and midrib somewhat to heavily thickened. Stem erect, unbranched, sheathed for most of its length, flexed sharply outward above the sheath of the upper leaf. Spike (2)4–9-flowered, inclined at an angle of about 30°, slightly flexuous; bracts green, sometimes flushed purple, 6–9 mm long, generally slightly more than one internode long, thus usually overlapping the base of the next bract, the inner slightly shorter than the outer. Flowers pale-yellow or sometimes whitish or pink, the keels sometimes red to brown, the lower 3 tepals each with a deep-yellow mark in the upper half and this often with a greenish or brown outline; perianth tube 3.5–4.5 mm long, curved outward and emerging from between the bracts; tepals unequal, the uppermost larger than the others and hooded over the stamens, 8–10 × c. 6 mm, the upper lateral tepals directed forward, the lower 3 tepals 5–7 × 1.5 mm at the widest, much exceeding the upper when viewed in profile, joined to the upper lateral tepals for 3 mm and to themselves for c. 2 mm, closely aligned and more or less horizontal, abruptly narrowed below into claws. Filaments c. 5 mm long, unilateral and arcuate, exserted c. 2.5 mm from the tube; anthers yellow, parallel, 4.5–5 mm long, shortly apiculate, reaching or barely exceeding the apices of the upper lateral tepals. Style dividing opposite the middle of the anthers, style branches c. 1 mm long, not reaching the anther apices. Capsules 5–6 mm long, broadly obovoid.

Zambia. N: c. 8 km E of Kasama, 4.xii.1961, *Robinson* 4716 (K; PRE; SRGH). W: Chingola, 28.xi.1959, *Fanshawe* 5267 (K; NDO).

Restricted to northern Zambia and adjacent Zaire and W Tanzania. In seasonally wet habitats such as dambos and near seasonal pools and streams; flowering in November and December, at the end of the dry season.

The tiny yellow to white flowers with perianth tubes 3.5–4 mm long and the uppermost tepals 8–10 mm long distinguish *G. pusillus*. Until now it has been confused with *G. gracillimus* Baker as both species share a similar and unusual vegetative morphology which is characterised by a slender habit together with the fine fibrous corm tunics, the strongly flexed spike, and the possession of only 2 leaves the lower one of which sheaths most of the stem. The two species are readily differentiated from one another because *G. gracillimus* Baker has large pale-blue to lilac or occasionally dark-purple flowers with a perianth tube 11–15 mm long and uppermost tepals 11–15 mm long.

20. **Gladiolus erectiflorus** Baker in Bull. Misc. Inform., Kew **1895**: 293 (1895); in F.T.A. **7**: 367 (1898). —Geerinck in Bull. Jard. Bot. Belg. **42**: 274 (1972), excl. *G. linearifolius*. TAB. **25** fig. A. Type: Zambia, Liendwe, south of Lake Tanganyika, 1894, *Carson* 1/1894 (K, holotype).

Gladiolus nyikensis Baker in Bull. Misc. Inform., Kew **1897**: 283 (1897); in F.T.A. **7**: 367 (1898). Type: Malawi, "Nyika Plateau, 6000–7000 ft", *Whyte* s.n. (K, holotype; B; G).

Gladiolus venulosus Baker in Bull. Misc. Inform., Kew **1897**: 282 (1897); in F.T.A. **7**: 366 (1898). Type: Malawi, near Chitipa (Fort Hill), Tanganyika Plateau, 3500–4000 ft, vii.1896, *Whyte* s.n. (K, holotype; B; G).

Plants (35)60–90 cm high. Corms 12–18 mm in diameter; tunics of papery to matted fibrous layers, fragmenting irregularly or breaking into vertical strips below. Foliage leaves 5–6, the lower 3 more or less basal and larger than the others, reaching to about the base of the spike, 2–5 mm wide, narrowly lanceolate to nearly linear, the midrib and margins slightly thickened; the upper leaves cauline and decreasing in size above, imbricate, sheathing the stem almost to the base of the spike. Stem erect, unbranched. Spike 4–8-flowered, erect, rarely flexed at the base and weakly inclined; bracts green below, membranous above, becoming dry and light-brown toward the apices, 20–30(40) mm long, lanceolate-attenuate, the inner slightly shorter than the outer. Flowers cream-coloured to pale-pink, or red to purple, with conspicuous pink to red or purple veins, the venation often darker-coloured on fading (the flowers rarely apparently uniformly pale-pink to whitish), the lower lateral tepals or all 3 lower tepals each with a broad lanceolate cream-coloured to yellow mark in the lower centre midline sometimes with a pink streak above it; perianth tube 16–18 mm long, narrowly infundibuliform, widening and curving outward near the apex; tepals subequal or the uppermost slightly larger, (25)30–35 mm long (often less when dry), lanceolate, the uppermost inclined over the stamens, the lower 3 tepals united with the upper lateral tepals for c. 2 mm, straight, usually inclined weakly toward the ground. Filaments 12–14 mm long, ultimately exserted 6–7 mm; anthers 7–8 mm long, apices with minute subacute appendages less than 0.2 mm long. Style dividing at about the apex of the anthers, style branches c. 4 mm long, extending beyond the anthers. Capsules 10–12 mm long, obovoid-ellipsoid.

Zambia. N: Kambole Escarpment, dry rocks near Ngozye (Ngozi) Falls, 6.vi.1957, *Richards* 10055 (BR; K; SRGH). W: Kitwe, Parklands, c. 1250 m, 8.v.1961, *Linley* 144 (K; LISC; MO; SRGH). C: c. 8 km E of Chiwefwe, 15.vii.1930, *Hutchinson & Gillett* 3665 (BM; BR; K; SRGH). E: Katete, St. Francis' Hospital, 21.iv.1957, *Wright* 190 (BR; K). **Malawi**. N: Mzimba Distr. c. 30 km W of Mzuzu, Lunyangwa R. Bridge, 18.iv.1974, *Pawek* 8342 (BR; K; MAL; MO; SRGH; WAG). C: Chencherere Hill, N of Dedza, 24.iv.1971, *Pawek* 4701 (K; MAL; SRGH). **Mozambique**. N: c. 9 km W of Namina, c. 640 m, 23.v.1961, *Leach & Rutherford-Smith* 10981 (K; MO; SRGH).

Also in Zaire (Shaba) and SW Tanzania. In deciduous woodland, often in rocky sites in hilly country; flowering mostly in April and May, but sometimes as early as February and as late as July.

This species is striking in its moderate-sized flowers with subequal tepals which are white to pale-pink in colour and heavily veined in red to purple.

21. **Gladiolus verdickii** De Wild. & T. Durand in Bull. Soc. Roy. Bot. Belgique **40**: 29 (1901). Type from Zaire.

Gladiolus arnoldianus De Wild. & T. Durand in Bull. Soc. Roy. Bot. Belgique **40**: 27 (1901). Type from Zaire.

Plants 70–110 cm high. Corms 20–35 mm in diameter, bearing several broad fasciated stolons from the base, each with numerous terminal cormlets; tunics of papery to matted-fibrous layers, fragmenting irregularly or breaking into vertical fibres below. Foliage leaves 5–7, the lower 3–4 leaves more or less basal and larger than the others, reaching to about the base of the spike, (8)10–15 mm wide, narrowly lanceolate to linear, the midrib and margins hyaline and slightly thickened; the upper leaves cauline and decreasing in size above. Stem unbranched, rarely with 1 short branch. Spike 4–8-flowered, flexed at the base; bracts green below, membranous above and becoming dry and light-brown apically, (30)40–45(55) mm long, lanceolate-attenuate, the inner slightly shorter than the outer and forked apically. Flowers white to yellow, sometimes with conspicuous dark-red to pink veins, the lower lateral tepals each with a cream to yellow mark in the lower centre; perianth tube 25–35 mm long, narrowly funnel-shaped, widening and curving outward above; tepals unequal, lanceolate-oblong, the uppermost 45–50 mm long (often less when dry), larger than the others and inclined over the stamens, the upper lateral tepals slightly shorter and spreading at right angles to the tube, the lower lateral tepals smallest, 38–40 mm long. Filaments c. 30 mm long, exserted 14–18 mm; anthers 7–8 mm long, apices with vestigial appendages less than 0.2 mm long. Style dividing near the anther apices, style branches c. 3.5 mm long, spreading beyond the anthers. Capsules 10–16(20) mm long, obovoid-ellipsoid.

Zambia. W: Mwinilunga Distr., Solwezi road, 79 km east of Mwinilunga and 10 km northwest of Lumwana Mission, 14.v.1972, *Kornaś* 1785 (K).

Tab. 25. A. —GLADIOLUS ERECTIFLORUS, whole plant (×½), flower (×¾), from *Goldblatt* 9091.
B. —GLADIOLUS SERICEOVILLOSUS subsp. CALVATUS, whole plant (×⅓), front view
of flower (×¾), from *Goldblatt* 9083. Drawn by J.C. Manning.

Also in Zaire (Shaba), recorded in Zambia only once. In open woodland; flowering mostly in April and May, occasionally as late as July.

Similar to the related *G. erectiflorus* in the cream to pink tepals which are usually heavily veined in pink to purple, *G. verdickii* is distinguished by the larger flower with a perianth tube 25–35 mm long and tepals 45–50 mm long. The fasciated cormiferous basal stolons are unique in the genus.

22. **Gladiolus sericeovillosus** Hook. f. in Bot. Mag., t. 5427 (1864). Type is a plant cultivated at Kew from material from South Africa.

Subsp. **calvatus** (Baker) Goldblatt, comb. et stat. nov. TAB. **25** fig. B.

> *Gladiolus ochroleucus* sensu Baker in Bot. Mag. **103**: t. 6291 (1877) non Baker (1876).
> *Gladiolus ludwigii* var. *calvatus* Baker in F.C. **6**: 150 (1896). —Phillips in Fl. Pl. S. Africa **4**: t. 125 (1924).
> *Gladiolus sericeovillosus* var. *glabrescens* L. Bolus in S. African Gard. **18**: 213 (1928). Type from South Africa (Transvaal).
> *Gladiolus dehnianus* Merxm. in Proc. Trans. Rhodesia Sci. Assoc. **43**: 151 (1951). Type: Zimbabwe, Marondera (Marandellas), 30.iii.1941, *Dehn* 14 (M, holotype, not seen; K; SRGH).
> *Gladiolus elliotii* sensu G.J. Lewis et al. in J. S. African Bot., Suppl. 10: 27 (1972) pro parte, non Baker (1891). —Tredgold & Biegel, Rhod. Wild Fl.: 11, pl. 8 (1979).
> *Gladiolus sericeovillosus* forma *calvatus* (Baker) Oberm. in J. S. African Bot., Suppl. 10: 31 (1972).

Plants 35–100 cm high. Corms 2.5–3 cm in diameter, globose; tunics coriaceus, decaying into vertical fibrous strips. Foliage leaves 5–7, mostly basal, exceeding the spike sometimes by up to twice its length, 3–6(8) mm wide, linear (narrowly lanceolate in South Africa), usually (microscopically) pubescent on the sheaths and often sparsely so on the lower parts of the leaves, the margins and midrib and sometimes the other veins moderately to strongly thickened and hyaline; the upper 1–2 leaves cauline and much shorter than the basal. Stem rarely branched. Spike distichous, erect, 12–20(40)-flowered; bracts pale-green and soft-textured, becoming membranous or dry and pale above, 15–30 mm long, the outer acute-attenuate, usually slightly longer (or shorter) than the inner, the inner forked apically, the margins united around the flower. Flowers pale grey-green to cream-coloured, the tepals densely speckled with minute dark-red to maroon points, these often concentrated in the midline of each tepal, the lower lateral (and sometimes the lowermost) tepals yellow-green in the lower half (tepals sometimes uniformly cream with green markings on the lower tepals in South Africa); perianth tube (6)10–12 mm long, obliquely infundibuliform, widening and curving outwards near the apex, extended between the bracts; tepals unequal, the uppermost (18)20–22 mm long (often less when dry), longer than the others and hooded over the stamens, the lower 3 tepals c. 12 mm long, curving toward the ground, narrowed somewhat below but not abruptly expanded into limbs above. Filaments 10–12 mm long, exserted 6–8 mm; anthers 8–10 mm long. Style dividing near anther apices, style branches 4–5 mm long. Capsules c. 12 mm long, obovoid, usually somewhat trilobed above.

Zimbabwe. C: Harare, Cleveland Dam, 15.v.1948, *Wild* 2547 (K; SRGH). E: Mutare Distr., near Odzi, 11.v.1963, *Chase* 8016 (K; P; SRGH). S: Mberengwa (Belingwe) Mt., iv.1978, *Cannell* 739 (SRGH).

Also in South Africa (Transvaal). In open tall grassland or in light woodland in well drained soils; usually flowering in March and April.

This subspecies is distinguished by the distichous erect spike of moderate-sized, dull-coloured flowers and long linear leaves that normally exceed the spikes and have strongly thickened margins and midribs. Subspecies *sericeovillosus*, which occurs in Natal, E Cape Province and S Transvaal, has long silvery hairs on the bracts and spike axis, and leaves shorter than the stems.

23. **Gladiolus elliotii** Baker in J. Bot. **29**: 70 (1891); Handb. Irid.: 215 (1892); in F.C. **6**: 150 (1896). —G.J. Lewis et al. in J. S. African Bot., Suppl. 10: 27–29, pl. 5 (1972) excl. syn. *G. dehnianus* Merxm. —van Wyk & Malan, Wild Fl. Witwatersrand & Pretoria: 62, 218 (1988). Type from South Africa (Transvaal).

> *Gladiolus rigidifolius* Baker in Bull. Herb. Boissier, sér. 2, **4**: 1006 (1904). Type from South Africa (Transvaal).

Plants 35–80 cm high. Corms 2.0–2.5 cm in diameter; tunics membranous to coarsely fibrous. Foliage leaves 3–4 basal, longer than the others, reaching at least to the base of the spike, rarely shortly exceeding it, 10–15 mm wide, narrowly lanceolate (to linear), sometimes minutely pubescent, the margins thickened and hyaline, the midrib less so and some other veins equally prominent; the upper leaves deceasing progressively in size, the uppermost sometimes entirely sheathing. Stem erect, rarely branched. Spike erect, distichous, 9–15-flowered; bracts green or somewhat dry and light-brown above, 2–3 cm long, usually imbricate, often 1.5(2) internodes long, clasping the stem below, apices curved outwards, the inner about as long as the outer and bifid with acuminate-setaceous apices. Flowers white to pale-blue or purple, usually densely speckled with minute purple to pink dots, these more crowded towards the midline of the tepals, the lower tepals each with a central yellow streak in the lower half; perianth tube 15–16 mm long, shortly exserted from the bracts, narrowly funnel-shaped; tepals 25–33 mm long, lanceolate, the uppermost larger than the others and arching over the stamens, the lower 3 tepals united for 2–3 mm, narrower below, curving downward throughout, usually slightly exceeding the uppermost when viewed in profile. Filaments 8–10 mm long, exserted for 3–4 mm; anthers 8–11 mm long. Style dividing near the anther apices, style branches c. 5 mm long. Capsules c. 2.5 cm long, nearly ellipsoid.

Botswana. SE: Kanye, 1895, *Marloth* 2162 (PRE).
Widespread in eastern southern Africa. Locally in grassland (SE Botswana), usually above 1000 m, sometimes in seasonally wet sites; flowering in November to February.
This species is allied to, and often confused with, *G. sericeovillosus* subsp. *calvatus* which also has a dense distichous spike, but may be distinguished by the white to pale-bluish perianth densely speckled with pale-blue dots, and by the shorter leaves with several equally prominent veins and thick hyaline margins. The typically taller *G. sericeovillosus* subsp. *calvatus* has greenish flowers densely red-dotted and long linear leaves which exceed the spikes and in which only the midrib and margins are thickened.

24. **Gladiolus ecklonii** Lehm., Del. Sem. Hort. Hamburg: 7 (1835). Type from South Africa.

Subsp. **rehmannii** (Baker) Oberm. in J. S. African Bot., Suppl. 10: 44 (1972). Type from South Africa (Transvaal).
    Gladiolus rehmannii Baker, Handb. Irid.: 216 (1892); in F.C. **6**: 153 (1896). —Pole Evans in Fl. Pl. S. Africa **1**: t. 20 (1921). Type as above.
    Gladiolus cymbarius Baker in Bull. Herb. Boissier, sér. 2, **1**: 866 (1901). Type from South Africa (Transvaal).

Plants 35–100 cm high. Corms 1.5–5 cm in diameter; tunics of brown matted coarse fibres usually extending upwards into a neck 2–6 cm long; numerous basal sessile ovoid cormlets often present. Foliage leaves glaucous green, 6–10, reaching to at least the base of the spike, more usually to the middle of the spike, 6–15 mm wide, linear to narrowly lanceolate, margins and midrib hardly thickened, but differentiated from the other veins; upper cauline leaves 1–2, smaller than the basal, sometimes entirely sheathing. Stem erect, unbranched. Spike 6–14-flowered, erect; outer bracts glaucous green, 5–7(8) cm long, imbricate, 2–3 internodes long, the inner about half as long as the outer. Flowers mauve, pink or white, the lower lateral tepals with a yellow median blotch in the lower half, large, up to about 6 cm long, funnel-shaped, partly enclosed within the bracts; perianth tube 18–22 mm long, obliquely infundibuliform; tepals 45–55 mm long, oblong to broadly ovate, the uppermost ascending and slightly hooded, the lower lateral tepals smaller, curved outward. Filaments 10–12 mm, included or exserted 2–3 mm; anthers 12–15 mm long. Style dividing near the anther apices, ultimately exceeding them, style branches c. 6 mm long. Capsules 15–30 mm long, oblong to ellipsoid.

Botswana. SE: Kweneng Distr., 32 km from Molepolole-Letlhakeng road towards Ngware, 17.iii.1978, *Hansen* 3380 (C).
Also in South Africa (Transvaal). Rare and local in SE Botswana in rocky grassland and sandy flats; flowering mostly in March and April.
Subspecies *rehmannii* is distinguished by its long imbricate floral bracts, 5–7(8) cm long and at least 2 internodes in length, and by the linear or narrowly lanceolate leaves. The mauve, pink or white flowers of subsp. *rehmannii* lack the dark-red to maroon spots on the tepals characteristic of subsp. *ecklonii* and subsp. *vinosomaculatus*.
The leaves of subsp. *ecklonii* are broadly lanceolate (rarely narrowly so), about half as long as the stems, and are (11)15–30 mm wide with margins and midribs heavily thickened. Subspecies *rehmannii* is so different both in leaf and flower that it should probably be treated as a separate species. Subspecies *vinosomaculatus*, however, appears to link subsp. *rehmannii* and subsp. *ecklonii*.
Subspecies *ecklonii* is widespread in eastern southern Africa and Swaziland, but subsp. *vinosomaculatus* restricted to the southern and central Transvaal.

25. **Gladiolus dalenii** Van Geel, Sert. Bot., fasc. 28 (1829). —Hilliard & Burtt in Notes Roy. Bot. Gard. Edinburgh **37**: 297 (1979). —Goldblatt in Fl. Madagascar, fam. 45 (ed. 2), 42 (1991). TAB. **26**. Type from South Africa (Natal).

Tab. 26. GLADIOLUS DALENII, whole plant (×⅔), dissected flower (×1), from *Pawek* 6340. Drawn by J.C. Manning.

Plants (50)70–120(150) cm high. Corms (15)20–30 mm in diameter; tunics reddish-brown, of brittle or coriaceous to membranous layers, the outer becoming irregularly broken, sometimes fibrous; numerous tiny cormlets usually present around the base. Foliage leaves either synanthus and produced simultaneously with the flowers on the flowering shoot (subsp. *dalenii*) or hysteranthus and produced after flowering on separate shoots from the same corm (subsp. *andongensis*); synanthus foliage leaves 4–6(7), at least the lower 2 basal or nearly so, long-laminate and about half as long as the spike, (5)10–20(30) mm wide, narrowly lanceolate to nearly linear, firm textured with moderately raised thickened midribs and margins, the upper 1–2 leaves (of plants with synanthous leaves) cauline and sheathing for at least half their length, sometimes entirely sheathing, often imbricate; hysteranthus foliage leaves usually only 2 and resembling basal synanthus foliage leaves, flowering stem leaves (of plants with hysteranthus leaves) entirely sheathing or with blades up to 6 cm long. Stem unbranched. Spike (2)3–7(14)-flowered, erect; bracts green, sometimes dry and pale apically, (2.5)4–7 cm long, the inner slightly shorter than the outer. Flowers either red to orange with a yellow mark on each of the 3 lower tepals, or flowers yellow to greenish and often with red to brown streaks on the upper tepals; perianth tube (25)35–45 mm long, nearly cylindric and curving outward in the upper half; tepals unequal, the 3 upper tepals 35–50 × 22–30 mm and broadly elliptic-obovate, the uppermost larger than the others and horizontal to down-curved concealing the stamens, the upper lateral tepals about as long as, to c. 5 mm shorter than, the uppermost, 20–30 mm wide and directed forwards often curving outward distally, the lower 3 tepals 20–25(30) × 8–12 mm and curving downwards, the lowermost somewhat longer and narrower than the lower lateral tepals. Filaments c. 25 mm long, exserted 15–18 mm; anthers 12–16 mm long, pale-yellow. Style dividing near or beyond the anther apices, style branches (4)5–6 mm long. Capsules (18)25–35 mm long, ellipsoid to ovoid.

Gladiolus dalenii is easily recognised by the large flowers, 60–80 mm long with the upper 3 tepals 35–50 mm long and much exceeding the downcurved lower tepals, as well as the strongly hooded uppermost tepal, the well exserted anthers and the fairly long perianth tube which slightly exceeds the long floral bracts. The foliage leaves are typically broad and sword-shaped. It is widespread in sub-Saharan Africa and Madagascar and common throughout the Flora Zambesiaca region except for the semi-arid to arid parts of Botswana and southern Zimbabwe. It occurs in grasslands and woodlands and occasionally in dambos.

It is a striking ornamental plant, now widely cultivated. A late summer-flowering form from southern Africa with scarlet flowers is perhaps the best known to horticulture. More important than its value as a wild species in garden displays, is the role of *G. dalenii* as one of the parents in the original crosses that led to the development of the large-flowered *Gladiolus* hybrids which today are one of the worlds most important cut-flower crops.

G. dalenii is also one of the few species of *Gladiolus* useful to man other than as an ornamental. It is used in parts of West and southern Africa in preparations to cure both constipation and severe dysentery, and is 'highly esteemed in curing snake-bites' (fide *Irvine*). Elsewhere in tropical Africa the corms may be eaten; the starchy corms are boiled and then leached in water before they are eaten. In southern Africa *G. dalenii* (under several synonyms) is a common component of the herbalist's medicine horn (lenaka). Its uses in treating cases of diarrhoea and as a cold remedy are also documented (fide *Watt & Breyer-Brandwijk*, 1962; *Jacot-Guillarmod*, 1971).

There are records that, at least in West Africa, *G. dalenii* is cultivated on farms in the forest, where it was introduced from the savanna country to the north (fide *Dalziel*, 1937). How much of its remarkably wide distribution is due to deliberate human activity may never be known.

Almost throughout its range, *Gladiolus dalenii* occurs in populations with flowers either pale-yellow (*G. primulinus*) to greenish with or without red to brown streaking, or flowers orange to red with bright-yellow and sometimes green markings on the lower tepals. Shorter tubed yellow-flowered populations from western Zambia, eg. *Robinson 5471*, with 2–3 foliage leaves and short spikes of 2–3 flowers, fit awkwardly in *G. dalenii*, as do a few collections from Malawi with purple flowers (*la Croix 287, Richards 14462*). Too little is known about the latter series of populations, always found in wetlands. They may on further investigation be found to warrant taxonomic recognition.

Leaves of the flowering stem with long well developed blades, these leaves produced simultaneously with the flowers on the flowering shoot - - - - - - - subsp. *dalenii*
Leaves of the flowering stem entirely sheathing or with short blades up to 10(15) cm long, long-bladed foliage leaves produced after flowering on separate shoots from the same corm
subsp. *andongensis*

Subsp. **dalenii**

Watsonia natalensis Eckl., Topogr. Verz. Pflanzensamml. Ecklon: 34 (1827) non *Gladiolus natalensis* Hook.f. (1831). Type from South Africa (Natal).

Gladiolus psittacinus Hook.f. in Bot. Mag. **57**: t. 3032 (1830). —Baker in F.C. **6**: 158 (1896). —Gomes e Sousa, Subsid. Estudo Fl. Niassa Port.: 56 (1935). —Suessenguth & Merxmüller, Contrib. Fl. Marandellas Distr.: 76 (1951). —Martineau, Rhod. Wild Fl.: 19, pl. 3 (1954). —Hepper in F.W.T.A. ed. 2, **3**(1): 141 (1968). Type from South Africa (Natal).

Gladiolus natalensis Reinw. ex Hook.f. in Bot. Mag. **58**: sub t. 3084 (1831) as nom. nov. pro *G. psittacinus*, nom. illegit. superfl. pro *G. psittacinus*. —G.J. Lewis et al. in J. S. African Bot., Suppl. 10: 44–53 (1972) [cited in error as (Eckl.) Reinw. ex Hook.f.]. —Geerinck in Bull. Jard. Bot. Belg. **42**: 281 (1972), excl. var. *melleri* (Baker) Geerinck. —Wickens, Fl. Jebel Marra: 158 (1976). —Plowes & Drummond, Wild Fl. Rhodesia: pl. 39, 40 (1976). —Tredgold & Biegel, Rhod. Wild Fl.: 11, plate 8 (1979). Type from South Africa (Natal).

Gladiolus quartinianus A. Rich., Tent. Fl. Abyssinica **2**: 306 (1851). —Baker in F.T.A. **7**: 371 (1898). —Eyles in Trans. Roy. Soc. S. Africa **5**: 331 (1916). —Hutch. in F.W.T.A. ed. 1: 379 (1936). —Mogg in Macnae & Kalk, Nat. Hist. Inhaca Isl., Moçamb.: 143 (1958). Type from Ethiopia.

Gladiolus luteolus Klatt in Peters, Naturw. Reise Mossambique: 515 (1864). —Baker in F.T.A. **7**: 368 (1898). Type: Mozambique, Boror, iv.1846, *Peters* s.n. (B, holotype).

Gladiolus corneus Oliver in Trans. Linn. Soc. London, **29**: 155, t. 100 (1875). —Baker, Handb. Irid.: 222 (1892); in F.T.A. **7**: 365 (1898). Type from Tanzania.

Gladiolus saltatorum Baker in Trans. Linn. Soc. London. **29**: 155 (1875). Type from Tanzania.

Gladiolus newii Baker in J. Bot. **14**: 334 (1876); Handb. Irid.: 214 (1892). Type from Tanzania.

Gladiolus angolensis Welw. ex Baker in Trans. Linn. Soc. London, ser. 2, Bot. **1**: 269 (1878); Handb. Irid.: 213 (1892). Type from Angola.

Gladiolus sulphureus Baker in Trans. Linn. Soc. London, ser. 2, Bot. **2**: 350 (1887); in F.T.A. **7**: 370 (1898), nom. illegit. non Jacq. (1786–93). Type from Tanzania.

Gladiolus primulinus Baker in Gard. Chron., ser. 3, **8**: 122 (1890). —Hutch. & Dalziel, F.W.T.A. ed. 1, **2**: 379 (1936). Type from Tanzania.

Gladiolus kilimandscharicus Pax in Engler, Hochgebirgsfl. Afr.: 175 (1892). —Baker, Handb. Irid.: 214 (1892). Type from Tanzania.

Gladiolus splendidus Rendle in J. Linn. Soc., Bot. **30**: 406 (1895). —Baker in F.T.A. **7**: 369 (1898). Type from Tanzania.

Gladiolus taylorianus Rendle in J. Linn. Soc., Bot. **30**: 405 (1895). Type from Kenya.

Gladiolus affinis De Wild., Pl. Nov. Horti. Then. **1**: 161, t. 35 (1905). Type: Mozambique, Morrumbala, xii.1900, *Luja* 393 (BR, holotype).

Gladiolus hockii De Wild. in Bull. Jard. Bot. État. **3**: 264 (1911). Type from Zaire (Shaba) probably subsp. *dalenii* but without stem base.

Gladiolus luembensis De Wild. in Bull. Jard. Bot. État **3**: 264 (1911). Type from Zaire (Shaba).

Gladiolus calothyrsus Vaupel in Bot. Jahrb. Syst. **48**: 537 (1913). Type from Tanzania.

Gladiolus boehmii Vaupel in Notizbl. Bot. Gart. Berlin-Dahlem **7**, number 68: 32 (1920). Type from Tanzania.

Gladiolus coccineus L. Bolus in S. African Gard. **18**: 213 (1928). Type: Zimbabwe, Paulington (cult. Hort. Kirstenbosch), *Stokes* s.n. (BOL, holotype mounted on 4 sheets; K).

Gladiolus louwii L. Bolus in J. Bot. **67**: 132 (1929). Type from Kenya.

Gladiolus barnardii G.J. Lewis in S. African Gard. **22**: 204 (1932). Type: Zimbabwe, near Harare (cult. Cape Town), *Barnard* s.n. (BOL, holotype mounted on 3 sheets).

(Further synonymy for *G. dalenii*, based on plants from South Africa, is cited by G.J. Lewis et al., 1972.)

Plants (50)70–120 cm high. Corms 20–30 mm in diameter. Foliage leaves produced simultaneously with the flowers on the flowering stem, 4–6(7), at least the lower 2 basal or nearly so, about half as long as the spike, (5)10–20(30) mm wide, narrowly lanceolate to more or less linear, firm textured with moderately raised and thickened midrib and margins; the upper 1–2 leaves cauline and sheathing for at least half their length, sometimes entirely sheathing, often imbricate. Spike (2)3–7(10)-flowered; bracts green, (3.5)4–7 cm long. Flowers with the perianth tube 35–45 mm long and uppermost tepal (35)40–50 mm long. Filaments 25–30 mm long, exserted 15–20 mm; anthers 12–16 mm long.

Caprivi Strip: c. 30 km from Katimo to Linyanti, 26.xii.1958, *Killick & Leistner* 3111 (PRE; SRGH). **Botswana**. N: Tsantsarra Pan, Chobe National Park, 22.i.1978, *P.A. Smith* 2201 (PRE; SRGH). **Zambia**. B: Barotse Plain, c. 20 km NW of Mongu, 10.x.1962, *Robinson* 5471 (SRGH); Senanga, 31.i.1975, *Brummitt, Chisumpa & Polhill* 14192 (K; SRGH; WAG). N: Mbala Distr., Nkali (Kali) Dambo, c. 1500 m, 15.i.1955, *Richards* 4102 (BR; K). W: Kitwe, Garneton, 28.xii.1961, *Linley* 232 (MO; SRGH). C: Serenje Distr., Kundalila Falls, 17.xii.1967, *Simon & Williamson* 1416 (SRGH). E: Chipata Distr., Nzomane, 6.i.1959, *Robson* 1043 (BM; BR). **Zimbabwe**. N: Mazowe (Mazoe), iv.1906, *Eyles* 360 (BM). W: Nyamandhlovu Pasture Research Station, 16.i.1954, *Plowes* 1660 (MO; PRE; SRGH). E: Mutasa Distr., Honde Valley, 750 m, 10.vi.1962, *Plowes* 2252 (SRGH). **Malawi**. N: Misuku Hills, Matipa area, c. 1920 m, 27.xii.1977, *Pawek* 13409 (K; MAL; MO). C: Chongoni Forestry School, 17.i.1967, *Salubeni* 501 (MAL; SRGH). S: Zomba, Chitinje Dambo, 3.i.1985, *Salubeni & Tawakali* 3919 (K; MAL; MO). **Mozambique**. N: Nampula, 17.ii.1937, *Torre* 1215 (COI; LISC). Z: Milange, 550 m,

17.ii.1972, *Correia & Marques* 2707 (BR; LMU). GI: Xai Xai (Vila de João Belo), Chipenhe, Floresta de Chirindzeni, 9.vi.1960, *de Lemos & Balsinhas* 44 (BM; COI; K; LISC; SRGH). M: Bokissa, 25 km from Maputo, 22.iv.1981, *Jansen & Macuácua* 7697 (MO; WAG).

Widespread in sub-Saharan Africa and Madagascar and common throughout the Flora Zambesiaca region except for the semi-arid to arid parts of Botswana and southern Zimbabwe. Flowering usually at least five to six weeks after the onset of the rainy season, in December to May (June). The stems always have well developed foliage leaves at flowering time.

Subsp. **andongensis** (Baker) Goldblatt comb. et stat. nov.

Gladiolus andongensis Welw. ex Baker in Trans. Linn. Soc. London, ser. 2, Bot. **1**: 269 (1878); in Handb. Irid.: 221 (1892). Type from Angola.

Gladiolus goetzei Harms in Bot. Jahrb. Syst. **28**: 365 (1900). Type from Tanzania.

Gladiolus pauciflorus De Wild. in Fedde, Repert. Spec. Nov. Regni Veg. **12**: 297 (1913) nom. illegit. non *G. pauciflorus* Baker (1878). Type from Zaire.

Gladiolus mildbraedii Vaupel in Notizbl. Bot. Gart. Berlin-Dahlem **7**, number 68: 33 (1920). Type from Rwanda.

Plants 60–90 cm high. Corms 15–30 mm in diameter. Leaves on the flowering stem 2–4, short and entirely sheathing, 6–14 cm long, or sometimes with blades 2–3(5) cm long and 6–12 mm wide, imbricate and sheathing the lower half of the stem; foliage leaves hysteranthus, usually at least 2, produced on separate shoots from the same corm after flowering, 4–16 mm wide and narrowly lanceolate. Spike (2)3–9-flowered; bracts (25)40–55 mm long. Perianth tube 25–33(40) mm long; uppermost tepal 35–45 mm long and 22–25 mm at the widest. Filaments c. 25 mm long, exserted 15–18 mm from the tube; anthers 12–15 mm long.

Zimbabwe. N: Mazowe Distr., Mvurwi (Umvukwes), Ruorka Ranch, c. 1550 m, 16.xii.1952, *Wild* 3915 (MO; SRGH). **Malawi**. N: Nyika Plateau, near Lake Kaulime, 8.i.1974, *Pawek* 7901 (MO; P); Rumphi, 8.xii.1965, *Banda* 763 (MAL; SRGH). **Mozambique**. N: Lichinga Plateau, xii.1932, *Gomes e Sousa* 1071 (COI). Z: Maganja da Costa, 64 km from Vila de Maganja to Mocuba, c. 100 m, 9.i.1968, *Torre & Correia* 16983 (LISC).

Fairly widespread in tropical Africa and extending from Senegal to Ethiopia and southwards to Mozambique. Flowering usually 3–4 weeks after the first rains, in November to December in south tropical Africa. The flowering stems of subsp. *andongensis* bear sheathing leaves only; later in the season long bladed foliage leaves are produced on separate non-flowering shoots. A third subspecies of *G. dalenii*, subsp. *welwitschii*, restricted to SW Angola, also has foliage leaves produced after flowering on separate shoots. It may be distinguished by a dense neck of fibres around the base and on upper tepal shorter than the upper laterals.

26. **Gladiolus velutinus** De Wild. in Fedde, Repert. Spec. Nov. Regni Veg. **12**: 297 (1913). Type from Zaire.

Gladiolus katubensis De Wild. in Fedde, Repert. Spec. Nov. Regni Veg. **12**: 297 (1913). Type: Zambia, Kafubu (Katuba) Stream, 30.xii.1907, *Kassner* 2268 (BR, holotype; B; K; P; Z).

Gladiolus vallidissimus Vaupel in Notizbl. Bot. Gart. Berlin-Dahlem **7**, number 68: 34 (1920). Type from Tanzania.

Gladiolus xanthus L. Bolus in S. African Gard. **23**: 47 (1933). Type: Zambia, Ndola West, 25.i.1932, *Duff* 59/32 (BOL, lectotype here designated).

Plants 1–1.5 m high. Corms 5–7 mm in diameter, depressed-globose; tunics reddish, irregularly broken. Foliage leaves several, mostly basal or inserted in the lower part of the stem, reaching to about the base of the spike, (13)20–35 mm wide, lanceolate, sometimes narrowly so, the margins, midrib and usually at least one other pair of veins thickened and hyaline, the surface densely and minutely papillose to short-pubescent. Stem unbranched. Spike 6–10-flowered, the nodes 3–7 cm apart; bracts green, 4–6(7) cm long, the inner 1–2 cm shorter than the outer. Flowers pale-yellow with greenish highlights; perianth tube 30–40 mm long, curving outward above, gradually expanding toward the apex; tepals unequal, the upper 3 tepals c. 4 cm long, the uppermost c. 2 cm wide and hooded over the stamens, the upper lateral tepals directed forward and curving outward only near the apex, the lower 3 tepals half to one-third as long as the upper, curving downward from the base. Filaments 25–30 mm long, exserted at least 15 mm; anthers 14–17 mm long. Style arching over the stamens, dividing near the anther apices, style branches c. 5 mm long. Capsules unknown.

Zambia. N: Mbala Distr., Kawimbe, i.1954, *Nash* 15 (BM; P; SRGH). **Malawi.** N: Nyika Plateau, Lake Kaulime, 4.i.1969, *Richards* 10446 (K).

Extending from southern Zaire across Zambia to western Tanzania and southern Uganda. In wetland sites such as marshes, dambos and streambanks; flowering in December and in January in Zambia and Malawi.

Allied to *G. dalenii* but having larger pale lemon-yellow and greenish flowers, and distinctive leaves the entire surfaces of which are covered with a dense minute papillose indumentum. Both the midrib and secondary veins are typically heavily thickened and hyaline.

27. **Gladiolus nyasicus** Goldblatt sp. nov. Plantae 40–60 cm altae, cormis 15–20 mm diametro, foliis lanceolatis 5–10 mm latis, spica (2)3–7(12) florum, floribus flavis vel cremeis, tubo perianthii 18–25 mm longo, tepalo superiore 20–28 mm longo, filamentis 4–6 mm exsertis, antheris 8–12 mm longis. Typus: Tanzania, Lumecha Bridge, c. 21 km N of Songea, 3.i.1956, *Milne-Redhead & Taylor* 8033 (K, holotype; B; SRGH).

Plants 40–60 cm high. Corms 15–20 mm in diameter; tunics light-brown, of brittle membranous layers, the outer becoming irregularly broken or fibrous. Foliage leaves produced simultaneously with the flowers on the same shoot, 4–5, the lower 2–3 leaves basal or nearly so, about a third to half as long as the spike, 5–10 mm wide and narrowly lanceolate, the blades firm-textured with moderately raised and thickened midrib and margins; the upper 2–3 leaves cauline and largely or entirely sheathing, often imbricate. Stem unbranched. Spike (2)3–7(10)-flowered; bracts pale-green, evidently becoming dry above, 24–30(40) mm long, the inner somewhat shorter than the outer. Flowers pale-yellow to cream-coloured, the uppermost tepal sometimes flushed with red; perianth tube 18–25 mm long, nearly cylindric below, curving outward and expanded in the upper half; tepals unequal, the 3 upper tepals broadly elliptic-ovate, the uppermost (20)25–28 × c. 16 mm, larger than the others and hooded over the stamens, the upper lateral tepals about as long as to c. 5 mm shorter than the upper and c. 12 mm wide, directed forward and curving outward distally, the lower 3 tepals 12–16 × 7–10 mm and down-curved, the lowermost somewhat longer than the lower laterals. Filaments c. 16 mm long, exserted 4–6 mm; anthers 8–12 mm long. Style dividing near the anther apices, style branches c 4.5 mm long. Capsules (18)20–24 mm long, ellipsoid-obovoid.

Malawi. C: Dedza, base of Ciwawo (Ciwau) hill, Chongoni Forestry School, 1650 m, 4.ii.1959, *Robson* 1447 (K; LISC; PRE; SRGH).

Also in S Tanzania. In boggy grassland and dambos; flowering in mid December to mid February.

G. nyasicus may be distinguished from the related *G. dalenii* by its relatively modest stature and small pale-yellow to cream-coloured flowers with smaller, down-curved lower tepals and a perianth tube 18–25 mm long. *G. dalenii* is 60–120 cm high, has perianth tubes (25)30–40 mm long, and upper tepals 35–50 mm long, and is altogether a more robust species. In addition to its smaller flowers, *G. nyasicus* can be distinguished from species of similar general appearance by its fairly short stamens, exserted 4–6 mm from the tube. However, the stamens are not as short as in *G. melleri* where they are fully included within the tube.

28. **Gladiolus magnificus** (Harms) Goldblatt in Bull. Mus. Natl. Hist. Nat., B, Adansonia **11**: 426 (1989). Type from Angola.

Antholyza zambesiaca Baker, Handb. Irid.: 232 (1892); in F.T.A. **7**: 374 (1898) non *Gladiolus zambesiacus* Baker (1892). Type: Botswana, Leshumo (Leshumo) Valley, *Holub* s.n. (K, holotype).
Antholyza spectabilis Schinz in Mém. Herb. Boissier, No. 20: 13 (1900) non *Gladiolus spectabilis* Baker (1904). Type from Namibia.
Antholyza magnifica Harms in Warburg, Kunene-Samb. Exped. Baum: 201 (1903). Type as above.
Petamenes zambesiacus (Baker) N.E. Br. in Trans. Roy. Soc. S. Africa **20**: 227 (1932). —Sölch in Merxm., Prodr. Fl. SW. Afrika, fam. 155: 12 (1969). Type as for *Antholyza zambesiaca* Baker.
Chasmanthe spectabilis (Schinz) N.E. Br. in Trans. Roy. Soc. S. Africa **20**: 273 (1932).
Petamenes magnifica (Harms) R.C. Foster in Contrib. Gray. Herb., no. 114: 50 (1936).
Petamenes spectabilis (Schinz) E. Phillips in Bothalia **4**: 44 (1941).
Oenostachys zambesiacus (Baker) Goldblatt in J. S. African Bot. **37**: 443 (1971). —Drummond & Plowes, Wild Fl. Rhodesia: pl. 44 (1976). Type as for *Antholyza zambesiaca* Baker.

Plants 80–140 cm high. Corms 2–3 cm in diameter, depressed globose; tunics of fine wiry straw-coloured fibres. Foliage leaves 5–7, the lower 4–5 basal and longer than the others, about half as long as the stem and reaching almost to the base of the spike, 6–9 mm wide, narrowly lanceolate, midrib and margins slightly thickened, the margins hardly raised; the upper leaves progressively smaller and with shorter blades, or the blades lacking and leaves sheathing only. Stem unbranched. Spike 6–15-flowered; bracts green, 2.5–3(3.5) cm long, the inner somewhat shorter than the outer. Flowers bright-red with yellow markings on the lower tepals and in the throat, the lower 3 tepals at least sometimes each with a dark red-purple blotch in the midline; perianth tube 25–30 mm long, narrow below, expanded near the top of the bracts and curving outward and nearly horizontal, c. 4 mm wide at the mouth; tepals very unequal, the uppermost 35–42 × 18–22 mm, larger than the others and hooded horizontal, the upper lateral tepals 13–15 mm long and broadly lanceolate, directed forwards, the apices curving outwards, the lower 3 tepals c. 10 mm long, narrowly lanceolate, more or less horizontal, or the lowermost recurving. Filaments 30–35 mm long, exserted c. 20 mm; anthers 12–17 mm long. Style dividing shortly below the apex of the anthers, style branches c. 3 mm long. Capsules unknown.

Botswana. N: Seronga Road, 0.4 km W of Masoko Pan, 22.i.1978, *P.A. Smith* 2204 (K; SRGH). **Zambia**. B: 30 km S of Mulobezi to Sesheke, 30.i.1970, *Anton-Smith* s.n. (SRGH). **Zimbabwe**. W: Hwange (Wankie) Game Reserve, Gwai (Gwaai) corridor, 14.ii.1956, *Wild* 4724 (COI; K; LISC; PRE; SRGH).
Also in S Angola and N Namibia. In light woodland or forest clearings in sandy soil, often with tall grasses (Kalahari Sandveld); flowering in January and February, occasionally in March.
This species is readily distiguished by the enlarged uppermost tepal which is 35–42 mm long and about twice as long as the upper lateral tepals, and by the lower 3 tepals which are much reduced in size and somewhat less than a third as long as the uppermost.

29. **Gladiolus bellus** C.H. Wright in Bull. Misc. Inform., Kew **1906**: 169 (1906). TAB. **27**. Type: Malawi, Mulanje (Mlanje), Tuchila Plateau, 6000 ft, v.1901, *Purves* 4 (K, lectotype).

Plants 60–90 cm high. Corms c. 15 mm in diameter; tunics reddish-brown, of brittle papery layers, the outer becoming irregularly broken, rarely subfibrous. Foliage leaves 4–5, the lower 3–4 basal or nearly so and reaching to at least the base of the spike, sometimes shortly exceeding it, 7–12 mm wide, narrowly lanceolate to more or less linear, firm textured with the midrib and margins slightly thickened; the upper 1–2 leaves cauline and much shorter than the basal. Stem usually inclined or drooping, unbranched. Spike 3–8-flowered; bracts green, becoming dry and brown above, 4–5 cm long, the inner slightly shorter than the outer. Flowers white, the lower tepals each with a dark-red to violet spade-shaped mark in the lower midline and usually with dark streaks in the throat; perianth tube (4.2)5–9 cm long, cylindric below, the upper part curved outwards and flared, c. 12 mm long; tepals subequal, 3–3.5 × 2–2.2 cm, broadly obovate. Filaments 22–28 mm long, exserted for 5–8 mm; anthers 9–12 mm long, apices with obscure appendages less than 0.2 mm long. Style dividing just beyond the anther apices, style branches c. 4.5 mm long. Capsules 30–35 mm long, oblong-ellipsoid.

Malawi. S: Mt. Mulanje, path from Tuchila Hut to head of Ruo basin, 6.iv.1970, *Brummitt* 9647 (K; MAL; PRE; SRGH).
Apparently endemic to Mt. Mulanje in S Malawi. In rocky sites as well as in wet grassland, mainly above 1800 m. An early record of this species from Mt. Zomba may be incorrect. It has not been found again in this well-collected area. Flowering from late March until June, occasionally into July.
G. bellus is distinguished by its elongate perianth tube, 4.2–8.8 cm long and about half as long again to twice as long as the bracts, and by its white perianth with large conspicuous red to violet markings on the lower tepals. White flowered *G. callianthus*, which has an even longer perianth, may be distinguished from *G. bellus* by its apiculate anther appendages (lacking in *G. bellus*) and tube (9)12–15 cm long.

30. **Gladiolus benguellensis** Baker in Trans. Linn. Soc. London, ser. 2, Bot. **1**: 268 (1878); Handb. Irid.: 221 (1892); in F.T.A. **7**: 370 (1898). Type from Angola.
   Tritonia tigrina Pax in Bot. Jahrb. Syst. **15**: 152 (1893). Type from Angola.
   Gladiolus pubescens Pax in Bot. Jahrb. Syst. **15**: 154 (1893). —Baker in F.T.A. **7**: 264 (1898), nom. illegit. non Lamarck (1791) [= *Babiana pubescens* (Lamarck) G.J. Lewis]; non Baker (1876) [= *Gladiolus pubigerus* G.J. Lewis].
   Gladiolus paxii Klatt in T. Durand & Schinz, Consp. Fl. Afric. **5**: 222 (1893) nom. nov. pro *Gladiolus pubescens* Pax. Type from Angola.
   Gladiolus macrophlebius Baker in F.T.A. **7**: 576 (1898). Type from Angola.
   Gladiolus malangensis Baker in Bull. Herb. Boissier, sér. 2, **1**: 867 (1901) nom. illegit. superfl. (based on the duplicate at Z of the type of *G. paxii*).
   Gladiolus longanus Harms in Warburg, Kunene-Samb.-Exped. Baum: 201 (1903). Type from Angola.
   Gladiolus pubescifolius G.J. Lewis in Ann. S. African Mus., **40**: 132 (1954) nom. illegit. superfl. (proposed as a new name for *Gladiolus pubescens* Pax).

Tab. 27. GLADIOLUS BELLUS, whole plant (×⅔), dissected flower (× 1), from *Chapman & Chapman* 7363. Drawn by J.C. Manning.

Plants 35–60 cm high. Corms dark-red, c. 10 mm in diameter, globose to horizontally elongate, often rhizome-like, 15–25 mm long, the old corms sometimes persistent and then attached horizontally; tunics reddish-brown, membranous, fragmenting irregularly. Foliage leaves 4–6, the basal longer than the others, usually with blades well-developed and reaching to about the base of the spike, 5–9 mm wide, narrowly lanceolate, glabrous to pubescent, sometimes with a dense appressed short pubescence, the margins and midribs moderately to heavily thickened and hyaline, sometimes other veins also somewhat thickened; the upper 1–2 cauline leaves usually entirely sheathing. Stem unbranched, erect. Spike (3)5–8-flowered, erect; bracts green, (20)25–30 mm long, often somewhat attenuate and dry apically, the inner about two-thirds as long as the outer. Flowers either scarlet-red with the lower tepals each with a median white stripe, or occasionally entirely yellow; perianth tube c. 15 mm long, expanding uniformly from base to apex, curving outward in the upper half; tepals unequal, the uppermost 18–21 × c. 12 mm, larger than the others and ascending to nearly horizontal, the upper lateral tepals c. 20 mm long, the lower 3 tepals 16–18 × c. 6 mm, more or less straight and held at 45° to the ground. Filaments 5–8 mm long, included or barely exserted; anthers 7–8 mm long. Style dividing opposite the upper half of the anthers, style branches 3–4 mm long. Capsules (14)16–18(22) mm long, narrowly obovoid to ellipsoid.

Zambia. W: Mwinilunga Distr., swamp 3 km from Kabompo Gorge, 24.xi.1962, *Richards* 17499 (K; SRGH).

Common in Angola and S Zaire. Evidently rare in W Zambia and there restricted to permanent or seasonal (drying out between rainy seasons) marshy sites; flowering late in the dry season, from September into the early part of the wet season in December and January.

G. benguellensis is recognised by its fairly small flowers with subequal tepals, a short perianth tube and short stamens, the filaments being only 5–8 mm long are usually included or just barely exserted from the tube. It is similar to *Gladiolus melleri* but the latter has larger flowers and nearly always lacks foliage leaves at flowering time.

31. **Gladiolus melleri** Baker in J. Bot. **14**: 334 (1876); Handb. Irid.: 212 (1892); in F.T.A. **7**: 362 (1898).
—Eyles in Trans. Roy. Soc. S. Africa **5**: 331 (1916). —Suessenguth & Merxmüller, Contrib. Fl. Marandellas Distr.: 76 (1951). —Plowes & Drummond, Wild Fl. Rhodesia: pl. 38 (1976). —Tredgold & Biegel, Rhod. Wild Fl.: 11, pl. 8 (1979). TAB. **28** fig. A. Type: Malawi, Mpimbe Hill, Shire R., x.1859, *Kirk* s.n. (K, lectotype here designated).
   Gladiolus brachyandrus Baker in Bot. Mag. **105**: t. 6463 (1879); Handb. Irid.: 214 (1892); in F.T.A. **7**: 370 (1898). Type: a specimen cultivated at the Botanic Gardens, Edinburgh from material collected by *Buchanan* s.n. in Malawi, Shire Highlands (consisting of the figure in Bot. Mag. t. 6463 - no preserved specimens found at E or K).
   Gladiolus welwitschii subsp. *brevispathus* Pax in Bot. Jahrb. Syst. **15**: 155 (1893). —Fries, Wiss. Ergebn. Schwed. Rhod.-Kongo-Exped.: 235 (1916). Type from Angola.
   Gladiolus brevispathus (Pax) Klatt in T. Durand & Schinz, Consp. Fl. Afric. **5**: 214 (1893). Type as for *G. welwitschii* subsp. *brevispathus* Pax.
   Gladiolus johnstonii Baker in Bull. Misc. Inform., Kew **1897**: 283 (1897); in F.T.A. **7**: 372 (1898). Type: Malawi, Mt. Zomba and vicinity, "2500–3500 ft.," xii.1896, *Whyte* s.n. (K, lectotype here designated).
   Gladiolus natalensis var. *melleri* (Baker) Geerinck in Bull. Jard. Bot. Belg. **42**: 285 (1972). Type as above.
   Gladiolus dalenii var. *melleri* (Baker) Còrdova, Bull. Jard. Bot. Belg. **60**: 326 (1990). Type as above.

Plants 25–40(60) cm high. Corms 2.5–3.5 cm in diameter; tunics straw-coloured, coriaceous, fragmenting irregularly, the outer layer sometimes becoming fibrous. Leaves borne on the flowering stem usually partly to entirely sheathing, hardly differing from the cataphylls, (1)2–3(more), sometimes with short blades 2–5(10) cm long, often partly dry by anthesis, occasionally flowering stems with 1–2 long-bladed leaves; foliage leaves produced after the flowers and on separate shoots, evidently solitary, ultimately at least 30 × c. 1 cm, lanceolate, the margins and midrib moderately thickened and hyaline. Stem simple, rarely branched, 3–4 mm in diameter at the base of the spike. Spike 5–9(12)-flowered; bracts green, often flushed reddish and becoming membranous above, (15)20–28(35) mm long, the inner slightly shorter than the outer. Flowers orange-red to pink, rarely yellow to whitish, without markings; perianth tube (12)18–20(30) mm long, widening evenly from the base, gently curving outwards; tepals unequal, the upper 3 tepals 25–38 × 15–18 mm and strongly recurved distally, the uppermost horizontal or down-tilted, the upper lateral tepals slightly narrower and directed forwards, the lower tepals curving downwards, the lower lateral tepals c. 25 × 12–15 mm, the lowermost 30–35 mm long, often about as long as the upper. Filaments 8–10 mm long, included in the

Tab. 28. A. —GLADIOLUS MELLERI, flowering stem and vegetative shoot with foliage leaf (×½), front view of flower (×¾), from *Goldblatt et al.* 8153. B. —GLADIOLUS HUILLENSIS, whole plant (×½), detail of flowering and fruiting spikes (×¾), from *Milne-Redhead* 3111. Drawn by J.C. Manning.

upper part of the tube; anthers 10–12 mm long. Style dividing 2–3 mm beyond the apex of the anthers, style branches 6–8 mm long. Capsules 15–20(25) mm long, oblong to ellipsoid.

Zambia. N: Kambole-Mbala road, 15.ix.1960, *Richards* 13260 (BR; K). W: Ndola, 24.ix.1954, *Fanshawe* 1566 (BR; K; NDO). C: Kabwe (Broken Hill), 15.ix.1964, *Mutimushi* 944 (NDO; SRGH). S: Pemba, ix.1909, *Rogers* 8556 (BM; BOL; K; Z). **Zimbabwe.** N: Mazowe (Mazoe), ix.1906, *Eyles* 417 (BM; BOL; SRGH). W: Matopos Research Station, 9.x.1952, *Plowes* 1491 (K; LISC; SRGH). C: c. 15 km N of Marondera (Marandellas), 4.x.1953, *Wild* 4141 (K; LISC; MO; PRE; SRGH). E: Mutare (Umtali), commonage, 9.x.1953, *Chase* 5113 (BM; SRGH). S: between Masvingo (Fort Victoria) and Ndanga, 20.x.1930, *Fries, Norlindh & Weimarck* 2133 (BR; S). **Malawi.** N: Nyika Plateau, c. 20 km S of Chelinda, 19.x.1975, *Pawek* 10320 (K; MO; PRE; SRGH). C: Lilongwe Distr., Dzalanyama Forest Reserve, 3.x.1962, *Banda* 453 (MAL; SRGH). S: Bvumbwe, c. 1200 m, 14.x.1982, *la Croix* 349 (K). **Mozambique.** T: between Furancungo and Mualadze (Vila Gamito), 20.x.1943, *Torre* 6075 (LISC). MS: Manica, 4 km from Rotanda to Mavita, 30.x.1965, *Correia* 289 (LISC).

Also in Angola, Zaire (Shaba) and W Tanzania. Common in seasonally wet sites, such as the margins of dambos and in poorly drained savanna; flowering at the end of the dry season, from August to November. Flowering is stimulated by fire and signs of burning are common on specimens.

G. melleri is distinctive in its brick- to salmon-coloured flowers and in its flowering stems bearing only dry sheathing leaves. Laminate leaves are produced later in the season on separate shoots. In lacking foliage leaves at flowering time it resembles *G. dalenii* subsp. *andongensis*, from which it is readily distinguished by the short stamens with the filaments always included in the tube and by the somewhat smaller flowers and bracts.

32. **Gladiolus huillensis** (Welw. ex Baker) Goldblatt in Bull. Mus. Natl. Hist. Nat., B, Adansonia **11**: 426 (1989). TAB. **28** fig. B. Type from Angola.

Antholyza huillensis Welw. ex Baker in Trans. Linn. Soc. London, ser. 2, Bot. **1**: 270 (1878); Handb. Irid.: 392 (1892); in F.T.A. **7**: 374 (1898). Type as above.

Gladiolus subulatus Baker in F.T.A. **7**: 577 (1898). Type from Angola.

Antholyza degasparisiana Buscal. & Muschl. in Bot. Jahrb. Syst. **49**: 463 (1913). Type: Zambia, Kabwe (Broken Hill), date unknown, *Aosta* 157 (location of the type unknown).

Antholyza pubescens Vaupel in Notizbl. Bot. Gart. Berlin-Dahlem **7**, number 68: 31 (1920). Type: Zambia, ?Nyeneshi R. ("N'Yengeshi Spruit"), 28.xii.1907, *Kassner* 2211 (B, holotype; BM; HBG; K; P).

Petamenes huillensis (Welw.) N.E. Br. in Trans. Roy. Soc. S. Africa **20**: 276 (1932).

Petamenes degasparisiana (Buscal. & Muschl.) N.E. Br. in Trans. Roy. Soc. S. Africa **20**: 277 (1932). Type as for *Antholyza degasparisiana*.

Petamenes vaginifer Milne-Redhead in Hook. Icon. Pl. **35**: t. 3478 (1950). Type: Zambia, Mwinilunga Distr., E of Dobeka bridge, 5.xi.1937, *Milne-Redhead* 3111 (K, lectotype, designated by Goldblatt & de Vos, 1989; BR; K).

Oenostachys vaginifer (Milne-Redhead) Goldblatt in J. S. African Bot. **37**: 443 (1971). — Geerinck in Bull. Soc. Roy. Bot. Belgique **105**: 7 (1972). Type as for *Petamenes vaginifer.*

Oenostachys huillensis (Welw. ex Baker) Goldblatt in J. S. African Bot. **37**: 443 (1971). Type from Angola.

Plants 30–100 cm high. Corms 16–20 mm in diameter; tunics firmly membranous to papery, fragmenting irregularly into mostly vertical strips. Leaves 3–5 on flowering stem, short and sheathing for most of their length, obscurely to obviously white pubescent, imbricate, usually at least 2 with short blades to 8 cm long; blades more or less linear, the margins and midrib thickened (well developed foliage leaves which are produced from separate shoots after flowering are not known and are probably not produced). Stem unbranched, sheathed except for a short distance below the spike, often with a short membranous sheathing leaf inserted on the upper third of the stem just above the uppermost sheathing leaf. Spike 6–14(20)-flowered; bracts green below, dry and light-brown above, 13–20(25) mm long, sometimes sparsely pubescent above, the inner about as long as or only slightly shorter than the outer. Flowers bright-red, often marked with yellow on the lower tepals and in the throat; perianth tube slender below and abruptly expanded into a cylindric upper part, the lower part 11–16 mm long and c. 1 mm in diameter, exserted from the bracts, the upper part 10–12 mm long and c. 3 mm in diameter, ascending to horizontal; tepals unequal, the uppermost c. 15 × 15 mm, spathulate, larger than the others and extending horizontally over the stamens, the upper lateral tepals deltoid, 5–6 mm long, joined to the uppermost for c. 3 mm, the lower 3 tepals lanceolate and directed forwards, the lower lateral tepals c. 6 mm long, reaching the apices of the upper lateral tepals, the lowermost tepal shorter than the rest, c. 3 mm long. Filaments c. 20 mm long, reaching to about the middle of the uppermost tepal; anthers c. 7 mm long, reaching nearly to the apex of the uppermost tepal or barely exceeding it.

Style dividing opposite the upper half of the anthers, style branches 3–3.5 mm long, ultimately reaching the anther apices. Capsules 10–16 mm long, broadly ovoid.

Zambia. B: Kataba, 12.xii.1960, *Fanshawe* 5966 (K; NDO; SRGH). C: Mumbwa Distr., Kafue National Park, Ngoma, 23.xii.1964, *Mitchell 25/87* (K; LISC).

Also in western Angola and S Zaire. Evidently fairly rare in central and northwestern Zambia. In light woodland; flowering in November and December, early in the season, before the leaf canopy is closed and the undergrowth thick and tall.

G. *huillensis* is distinguished by its leaves with short to vestigial blades and sparsely pubescent sheaths, together with the bright-red perianth, and the unusual form of the flowers. The upper tepal is extended horizontally much exceeding the others, the lateral tepals are only about half as long as the uppermost, and the 3 lower tepals are vestigial, only 3–6 mm long.

33. **Gladiolus decoratus** Baker in J. Bot. **14**: 333 (1876); Handb. Irid.: 222 (1892); in F.T.A. **7**: 370 (1898). Type: Mozambique, Morrumbala Mountain, "from the foot to 2000 ft.", xii.1858, *Kirk* s.n. (K, lectotype here designated); Morrumbala, in 1863 and 1866, *Kirk* s.n. (K; MO, syntypes).

 Gladiolus kirkii Baker, Handb. Irid.: 222 (1892) nom. illegit. non Baker (1890). Type from Zanzibar.

 Gladiolus zanguebaricus Baker in Bull. Misc. Inform., Kew **1897**: 282 (1897), in obs.; in F.T.A. **7**: 365 (1898), as nom. nov. pro *Gladiolus kirkii* Baker.

 Gladiolus quilimanensis Baker in F.T.A. **7**: 577 (1898). Type: Mozambique, Zambezia, Quelimane, 10.ii.1889, *Stuhlmann s.n.*(B, holotype).

 Gladiolus morumbalaensis De Wild. in Pl. Nov. Horti. Then. **1**: 17–20, t. 5 (1904). Type: Mozambique, Zambezia, Morrumbala, xii.1900, *Luja 393* (BR, lectotype, here designated; BR).

Plants 45–80 cm high. Corms 15–25 mm in diameter; tunics initially membranous, becoming more or less fibrous and matted with age, fibres brown, enclosing numerous small cormlets. Foliage leaves 4–5, the lower 3 more or less basal and longer than the others, reaching to about the base of the spike, 8–18(24) mm wide, lanceolate to linear, the upper 2 leaves cauline and sheathing in the lower half, fairly soft-textured, the midrib and margins not thickened. Stem unbranched. Spike 5–9-flowered; bracts green and soft-textured, the outer sometimes flushed red, 25–40(50) mm long, the inner two-thirds as long as the outer. Flowers bright-red, the lower tepals each with a median white to yellow streak, the marks on the lower lateral tepals spathulate and widest near the tepal apices; perianth tube c. 35 mm long, the lower part 18–25 mm long, slender and cylindric, the upper part 8–10 mm long, funnel-shaped; tepals unequal, lanceolate, the uppermost 45–55 × 15–20 mm, larger than the others and arched over the stamens, the lower 3 tepals united for 10–12 mm, c. 40 mm long, nearly horizontal or dipping towards the ground, extending as far outward as the upper or slightly further. Filaments 35–40 mm long, exserted 20–25 mm; anthers 10–12 mm long, reaching to about the upper third of the uppermost tepal, with a slender often inconspicuous apiculum 0.5–1 mm long. Style dividing opposite to or beyond the anther apices, style branches c. 4 mm long, wider and channelled above. Capsules 18–27 mm long, ellipsoid.

Malawi. S: Zomba, Likangala R., 2.i.1957, *Banda* 338 (BM; MAL; SRGH). **Mozambique**. N: Pemba (Porto Amélia) to Ancuabe, 200 m, 21.xii.1963, *Torre & Paiva* 9635 (LISC). Z: Murrupula, Serra Marrutulo, 13.i.1961, *Carvalho* 429 (K; LMU). MS: 40 km N of Dondo, Inhaminga Road, 120 m, 3.xii.1971, *Pope & Müller* 503 (K; SRGH).

Also in south and central Tanzania. Occurring locally in southern Malawi and central Mozambique. Usually in sandy soil, in woodland and forest, also on rock outcrops and rocky slopes, from close to sea level to 2000 m; flowering in December and January.

G. *decoratus* is distinctive in its scarlet flowers with pale spathulate markings on the lower lateral tepals, in the arched uppermost tepal, and in the stamens with well exserted filaments and apiculate anthers. It is easily confused with G. *oligophlebius* which has a pink perianth in which the lower lateral tepals usually have linear markings. This latter species also has shorter stamens, and the upper tepals are hardly arched, so that the perianth does not have the inflated appearance characteristic of G. *decoratus*.

34. **Gladiolus oligophlebius** Baker in Bull. Misc. Inform., Kew **1895**: 73 (1895); in F.T.A. **7**: 367 (1898). —Hepper in F.W.T.A. ed. 2, **3**(1): 144 (1968). Type: Zambia, Mbala (Abercorn), 1893, *Carson* 25 (K, holotype).

 Gladiolus caudatus Baker in Bull. Misc. Inform., Kew **1895**: 74 (1895); in F.T.A. **7**: 367 (1898). Type: Zambia, Mbala (Abercorn), *Carson* 19, Feb. 1893, (holotype, K - plants depauperate and poorly pressed but clearly conspecific with G. *oligophlebius*).

Plants 40–80 cm high. Corms 15–25 mm in diameter; tunics initially membranous, becoming more or less fibrous and matted with age, fibres brown, enclosing numerous small cormlets. Foliage leaves 4–5, fairly soft-textured, the midrib and margins not thickened, the lower 3 more or less basal and longer than the others, reaching to about the base of the spike, 8–18(24) mm wide, lanceolate to linear, the upper 2 leaves cauline and sheathing in the lower half. Stem unbranced. Spike (2)5–9-flowered; bracts pale-green and soft-textured, the outer sometimes flushed red, 25–40(50) mm long, the inner two-thirds as long as the outer. Flowers pale to deep-pink, the lower tepals each with a median white to yellow streak outlined in dark-red or purple, the mark broadest in the centre and largest on the lower lateral tepals; perianth tube 30–40 mm long, narrowly funnel-shaped, the lower part c. 20 mm long; tepals subequal or the uppermost somewhat larger than the others, 38–45 mm long, lanceolate, the uppermost tepal more or less horizontal, 15–20 mm wide, the lower 3 tepals joined for c. 5 mm, dipping slightly below horizontal, 8–12 mm wide. Filaments 15–20 mm long, included or exserted 10(15) mm from the tube; anthers 10–15 mm long, reaching to about the middle of the uppermost tepal, with an acute apiculus c. 1.5 mm long. Style dividing at or beyond the anther apices, style branches c. 4 mm long. Capsules c. 15 mm long, more or less elliptic.

Zambia. N: Kalambo Falls, 8.ii.1965, *Richards* 19605 (K; MO). **Malawi**. N: Nkhata Bay Distr., c. 8 km W of Chintheche, 500 m, 29.xii.1978, *Phillips* 4487 (K; MO; WAG; Z). C: Salima, close to the lake shore, 1.i.1963, *Chapman* 1782 (K; PRE; SRGH).
Also in W Tanzania and Zaire (Kasai). Occurring locally in N and C Malawi and N Zambia. Mostly in rocky sites, in open woodland or sometimes in grassland at low elevations; flowering mostly in December and January.
G. oligophlebius is closely related to, and is easily confused with *G. decoratus*. The latter is generally distinguished by the scarlet, inflated perianth, the strongly arched uppermost tepal and the longer stamens with filaments exserted for 15–20 mm.
A specimen from Mporokoso in Zambia (*Bredo* 5919) has leaf blades with short-scabrid margins and vein edges, and rather densely scabrid leaf sheaths and cataphylls. In other respects the plant seems to agree with *G. oligophlebius*, although the flowers are poorly preserved.

35. **Gladiolus serenjensis** Goldblatt sp. nov. Plantae 15–25(40) cm altae, foliis 5–6, basalibus linearibus marginibus costistisque non incrassatis vel hyalinis, spicis 3–6-floris, bracteis viridibus attenuatis imbricatis 15–20(25) mm longis, floribus roseis, tubo perianthii c. 15 mm longo parte inferiore 6 mm longa, tepalis subaequalibus obtusis 18–20 mm longis, filamentis c. 11 mm longis breviter e tubo exsertis, antheris c. 6.5 mm longis. Typus: Zambia, Serenje Distr., Kundalila Falls, crevices in sandstone, 1500 m, 4.ii.1973, *Strid* 2834 (C, holotype; K; MO).

Plants 15–25(40) cm high. Corms 14–25 mm in diameter; tunics of coarse fibres. Foliage leaves 5–6, the lower 3–4 basal and longer than the others, usually slightly exceeding the spike, (2)3–4 mm wide, linear (narrowly lanceolate), the margins and midrib not thickened, the upper 2 leaves cauline narrower and shorter than the basal leaves. Stems unbranched. Spike 3–6-flowered, internodes 10–15 mm long; bracts green, sometimes flushed purple above, 15–20(25) mm long, attenuate, imbricate, 1.5–2 internodes in length, the inner about half as long as the outer. Flowers pink, the lower tepals evidently without markings; perianth tube c. 15 mm long, obliquely and narrowly funnel-shaped, the lower cylindrical part 6 mm long, curving outwards between the bracts; tepals subequal, 18–20 × c. 8 mm, lanceolate, all similarly oriented, directed forwards and curving outwards distally. Filaments c. 11 mm long, usually shortly exserted from the mouth of the tube; anthers c. 6.5 mm long, unilateral, with short acute apicula c. 0.4 mm long. Style dividing opposite the anther apices, style branches 4–5 mm long, extending beyond the anthers. Capsules unknown.

Zambia. C: c. 148 km S of Mpika, 2.iii.1962, *Robinson* 4974 (K; SRGH).
Restricted to a small area of the Muchinga Mountains of central Zambia in the Serenje District. In hilly country, apparently on rock outcrops in thin soils or in crevices; flowering in December.
G. serenjensis is distinguished by its short stature, the rather soft-textured leaves and attractive bright-pink flowers with subequal unmarked tepals that extend forward for most of their length. The weakly apiculate anthers suggest that it is allied to *G. decoratus* and *G. oligophlebius* both of which have much larger flowers with pale markings on the lower lateral tepals.

36. **Gladiolus callianthus** Marais in Kew Bull. **28**: 311–317 (1973). Type as for *Acidanthera bicolor* Hochst.
Acidanthera bicolor Hochst. in Flora **27**: 25 (1844). —Baker, Handb. Irid.: 188 (1892); in F.T.A. **7**: 358 (1898), non *Gladiolus bicolor* Baker, (1877). Type from Ethiopia.
Ixia quartiniana A. Rich., Tent. Fl. Abyss. **2**: 310 (1851), non *Gladiolus quartinianus* A. Rich. (1851). Types from Ethiopia.

Sphaerospora gigantea Klatt in Linnaea **34**: 699 (1866), nom. illegit. superfl. pro *Ixia quartiniana*. Type as for *I. quartiniana*.

Plants 30–65 cm high. Corms 15–22 mm in diameter; tunics dark red-brown, firmly to softly membranous, fragmenting irregularly, sometimes becoming subfibrous. Foliage leaves 4–8, the lower 3–5 basal, reaching at least to the base of the spike, sometimes slightly exceeding it, 5–12 mm at the widest, narrowly lanceolate, relatively soft-textured and without thickened margins or midrib. Stem erect, unbranched. Spike 3–4-flowered; bracts green, 5–8(10) cm long, the inner much shorter and concealed by the outer. Flowers white, with a prominent dark-purple streak in the midline of all tepals or of the lower 3 only, sweetly scented particularly in the evenings; perianth tube (9)12–15 mm long, cylindric and nearly straight, slightly expanded in the throat; tepals subequal, 35–45 × 17–22 mm, lanceolate. Filaments exserted for 10–15 mm; anthers c. 15 mm long, with rigid filiform apicula 2–5 mm long. Style arching over the stamens, dividing beyond the anthers, style branches c. 5 mm long. Capsules 20–27 mm long, ellipsoid to narrowly ovoid.

Malawi. C: Dedza, Chencherere Hill, Chongoni Forest, 11.iii.1967, *Salubeni* 581 (K; LISC; MAL; PRE). S: Zomba Mt., near Chingwe's Hole, 7.ii.1985, *Salubeni & Nachamba* 4006 (K; MAL; MO). **Mozambique**. T: Angónia, N of Serra Dómuè, *Macuácua & Mateus* 1092 (B).

Also in Ethiopia and Tanzania. Occurring locally in the mountains of S and C Malawi and adjacent Mozambique. In rocky sites, often lightly shaded forest margins; flowering mostly in January to March in the Flora Zambesiaca region.

G. callianthus is recognised by its long-tubed white flowers splashed with dark-purple on the lower tepals, anthers with acute appendages 2–5 mm long, and well developed, soft-textured leaves. It is most closely related to the W African *G. aequinoctialis* Herb. which has firm textured leaves with thickened primary and secondary veins and rather shorter anther appendages, usually 1–2 mm long.

37. **Gladiolus curtifolius** Marais in Kew Bull. **28**: 313 (1973). Type as for *Acidanthera goetzei* Harms.
    Acidanthera goetzei Harms in Bot. Jahrb. Syst. **30**: 278 (1901) non *Gladiolus goetzei* Harms. Type from Tanzania.

Plants 25–35 cm high. Corms 12–18 mm in diameter; tunics brittle-papery, usually broken into vertical fibres below. Leaves 2–3, inserted on the lower half of the stem, almost entirely sheathing, 4–8 cm long, sometimes with short non-sheathing blades 1–3 cm long. Stem erect, unbranched, the lower half sheathed by the leaves, the upper half naked or sometimes with a short sheathing bract 2–3 cm below the first flower. Spike (1)2–4-flowered; bracts green or membranous, usually dry and brownish above at anthesis, 15–20 mm long, the inner as long as the outer, bifid. Flowers cream, the tepals symmetrically disposed, fragrant; perianth tube 5–7.5 cm long, cylindric, very slightly expanded in the upper third; tepals subequal, 18–22 mm long, lanceolate, the inner 3 slightly shorter than the outer. Filaments unilateral, exserted c. 2 mm from the tube; anthers c. 7 mm long. Style dividing close to the anther apices, style branches 3–4 mm long. Capsules 13–15 mm long, ellipsoid to narrowly obovoid.

Malawi. N: Mafinga Mts., NE slopes, 18.i.1964, *Robinson* 6294 (B; K; MAL; SRGH).

Also in SW Tanzania, locally endemic to the highlands around the northern end of Lake Malawi. In grassland or open woodland, generally above 1800 m; flowering in November and December.

G. curtifolius is recognised by its cream-coloured flowers with a well developed perianth tube and lightly pubescent, short-bladed leaves. The elongated perianth tube, 5–7.5 cm long, suggests a relationship with *G. callianthus* but it lacks purple markings on the lower tepals, and the unusual, long-apiculate anthers characteristic of that species and its allies.

INDEX TO BOTANICAL NAMES